PILLAR OF SAND

Other Norton/Worldwatch Books

Lester R. Brown et al.

State of the World 1984 through *1999* (an annual report on progress toward a sustainable society)

Vital Signs 1992 through *1999* (an annual report on the environmental trends that are shaping our future)

Environmental Alert Series

Lester R. Brown
Christopher Flavin
Sandra Postel
Saving the Planet

Alan Thein Durning
How Much Is Enough?

Sandra Postel
Last Oasis

Lester R. Brown
Hal Kane
Full House

Christopher Flavin
Nicholas Lenssen
Power Surge

Lester R. Brown
Who Will Feed China?

Lester R. Brown
Tough Choices

Michael Renner
Fighting for Survival

David Malin Roodman
The Natural Wealth of Nations

Chris Bright
Life Out of Bounds

Lester R. Brown
Gary Gardner
Brian Halweil
Beyond Malthus

PILLAR OF SAND

Can the Irrigation Miracle Last?

Sandra Postel

W•W• NORTON & COMPANY
New York   London

Copyright © 1999 by Worldwatch Institute
All rights reserved
Printed in the United States of America

First Edition

The WORLDWATCH INSTITUTE trademark is registered in the U.S. Patent and Trademark Office.

The text of this book is composed in Minion, with the display set in Baker Signet. Book design by Elizabeth Doherty. Composition by Worldwatch Institute; manufacturing by the Haddon Craftsmen, Inc.

ISBN 0-393-31937-7

W. W. Norton & Company, Inc.
500 Fifth Avenue, New York, N.Y. 10110
www.wwnorton.com

W. W. Norton & Company Ltd.
Castle House, 75/76 Wells Street, London W1T 3QT

2 3 4 5 6 7 8 9 0

♻ This book is printed on recycled paper.

CONTENTS

Contents

PREFACE

People often ask me how I became interested enough in water to devote my professional life to it. I've wondered if the answer might have something to do with growing up on Long Island, in eastern New York. Though rarely conscious of it, Long Islanders routinely experience the fundamental irony of water on Earth—scarcity amidst plenty. Surrounded by the sea, the island feels water-rich. I recall gazing across the Atlantic from my favorite beach as a small child and marveling at the thought that nothing but ocean lay between me and Portugal. And yet with no significant rivers and only groundwater to tap, fresh water on Long Island is limited and precious.

For nearly two decades now, I have tracked, analyzed, and written about issues pertaining to Earth's fresh water. Time as a consultant in California drummed home the difficulties posed by the serious mismatch between the location of people and available fresh water. But it wasn't until I joined the Worldwatch Institute in 1983 that I began to put a global picture of water problems together. That picture has focused and refocused in my mind over the years, but always the key challenge has been clear: How can we meet growing human needs for water without destroying the health of rivers, lakes, and other aquatic systems that we depend on and that provide so many benefits?

During the short span of 16 years, I have seen water problems spread from pockets of particular regions to vast areas on virtually every continent. My previous book, *Last Oasis*, published first in 1992 and reissued in 1997 in conjunction with a public television documentary of the same name, attempted to lay out the scale of the global water challenge in its many dimensions. *Last Oasis* also showed the large and largely untapped potential of conservation and efficiency to help move us toward a more sustainable relationship with fresh water.

With *Pillar of Sand*, I zero in on the most difficult challenge presented by water scarcity—growing enough food for our future population in a sustainable manner. As a casual student of history, I have always been struck by the role irrigation played in the rise and demise of civilizations. *Pillar of Sand* attempts to place our current set of challenges in a broad historical context and to demonstrate the danger of complacency when it comes to our agricultural foundation.

Early chapters look at the major ancient irrigation-based civilizations and the rise of our modern irrigation society. Against this backdrop, I survey the host of environmental, social, and political challenges now confronting irrigated agriculture—from water shortages to the growing competition for water between countries. The last third of the book sketches a blueprint for a more productive, ecologically sound, and potentially lasting form of irrigated agriculture. It describes a role for more advanced technologies—such as drip irrigation and computerized weather and soil moisture monitoring—but also for simple, inexpensive technologies that can provide access to irrigation for the tens of millions of poor farmers now lacking it. Chapter 10 examines the "rules of the game" and highlights the institutional reforms needed to move us in a positive direction.

Despite the daunting nature of the global water challenge, I remain optimistic about our ability to meet it. While traveling through parts of Bangladesh, India, China, Mexico, Israel, Egypt, South Africa, and Uzbekistan, as well as my native United States, I have seen firsthand so many promising projects and initiatives and met so many concerned and dedicated individuals that to be anything but optimistic feels a betrayal of the human spirit and of human ingenuity.

But my cautious side keeps rearing its head. So far, our degree of response to the challenge pales in comparison to its scale. With the lives and livelihoods of hundreds of millions of people at stake, not to mention the health and functioning of our aquatic environment, there is little time to waste in crafting a more productive and sustainable form of irrigated agriculture—one that can last for centuries. It is a task well beyond the scope of engineers, because success will involve reshaping many facets of society—from our individual consumption patterns to national economic activities and regional politics.

My hope is that *Pillar of Sand* will provide inspiration to take up this challenge, as well as some useful ideas for successfully meeting it. I welcome your thoughts and comments.

Sandra Postel

Amherst, Massachusetts
March 1999

ACKNOWLEDGMENTS

As the process of writing this book draws to a close, I am keenly aware of the many debts I owe to people, places, and institutions.

First and foremost, I am grateful to the Pew Charitable Trusts and specifically to the Pew Fellows Program in Conservation and the Environment for awarding me a generous fellowship that provided much of the financial support for my work during the last three years. This award not only gave me the freedom to undertake the research that has culminated in this book, it also linked me to a stellar group of other Pew Fellows who provided inspiration and intellectual support. I also thank the Colorado-based Land and Water Fund, especially its former executive director Brian Hanson, for offering to serve as my host institution during the Pew Fellowship.

They say you can't go home again, but I did just that in joining forces with the Worldwatch Institute for the publication of *Pillar of Sand*. Having formally left the institute five years ago, I see more clearly than ever how invaluable the work of Worldwatch is to the overall cause of creating a sustainable society. As a senior fellow of Worldwatch, I continue to benefit from the extensive information network and unique interdisciplinary perspective that makes Worldwatch one of the world's most influential research organizations. For this continuing affiliation, I thank Lester Brown, the entire Worldwatch staff, and the many foundations and individual donors who support the institute's work. Thanks in particular to the Wallace Genetic Foundation for support for *Pillar of Sand* and to Jean Wallace Douglas for her long-standing support for sustainable agriculture.

I am grateful to freelance editor Linda Starke, who has probably edited more environmental books than any other individual, for her steady hand and guidance throughout the editing and production process. Anyone who has worked with Linda knows the great feeling of comfort she gives an author because of her impressive degree of organization, unbeatable attention to detail, and ability to keep everything on schedule. Her firm but diplomatic style is ideal for ironing out problems with thin-skinned authors.

Elizabeth Doherty, the designer of *World Watch* magazine, voluntarily took on the challenge of designing *Pillar of Sand*. I could not have been more pleased with the outcome. Liz not only added this project to an incredibly overloaded work schedule, she accomplished it with a striking degree of cheerfulness, efficiency, and professional competence.

Also at Worldwatch, Lori Brown and Anne Smith helped track down and provide me with research materials. Suzanne Clift provided helpful administrative support. Even before the book was finished, Reah Janise Kauffman was discussing contracts for foreign-language editions. And as the time approaches for outreach and marketing, I look forward to working with the highly effective Worldwatch communications team of Dick Bell and Mary Caron. Their behind-the-scenes efforts have much to do with the success of Worldwatch and its extensive reach.

In joining forces with Worldwatch, I also benefit from the institute's long-standing relationship with W.W. Norton & Company in New York. An added plus for me was finding out that the production and promotion of *Pillar of Sand* would be under the direction of Amy Cherry, senior editor at Norton— who also happens to be an old friend from high school. Although we haven't seen each other in 25 years, I rested easy knowing that the book was in such good hands. Thanks go to the production team of Nomi Victor and Andrew Marasia at Norton as well.

Every author reaches a point in the writing process where constructive criticism and feedback are needed to take a book further toward the desired outcome. I am especially grateful to seven people—all of them internationally recognized for their work in the water field—who took time out of very busy schedules to read all or part of *Pillar of Sand*. Their comments, suggestions, and insights markedly improved the book. Many thanks to Daniel Beard, senior vice president at National Audubon Society and former Commissioner of the U.S. Bureau of Reclamation; Peter Gleick, president of the Pacific Institute for Studies in Development, Environment and Security; Ruth Meinzen-Dick, research fellow with the International Food Policy Research Institute; Paul Polak, president of International Development Enterprises; David Seckler, director-general of the International Water Management Institute; Amy Vickers, president of Amy Vickers & Associates, Inc.; and Aaron Wolf, assistant professor of geography at Oregon State University.

Several Worldwatch colleagues—Dick Bell, Lester Brown, Christopher Flavin, Gary Gardner, and Brian Halweil—also took time to offer valuable and insightful reviews. Dick, in particular, pushed me to hone the writing, weed out the jargon, and generally make the manuscript more accessible to a broader audience. In doing so, he helped me achieve my goals for the book. Brian also pitched in during the final stages with some important data checking.

In addition, I benefited from an informal research affiliation with the Environmental Studies program at Mount Holyoke College. As part of the Five College Consortium in western Massachusetts, Mount Holyoke gave me access to vital library resources. Most important, the link brought me the able research assistance of Anna Marriott, formerly a student at Mount Holyoke and now a budding television journalist in Glasgow, Scotland. Anna's help in tracking down and sorting through many books and articles for the historical material in Chapter 2 proved invaluable.

People too numerous to name have generously offered their time and hospitality during my research travels in recent years. Special thanks to Paul Polak and the staff of International Development Enterprises—especially Len Jornlin, Guru Naik, and Sudarshan Suryawanshi—for making my 1998 trip to Bangladesh and India so fruitful and enjoyable. Among the others I would like to thank are Urs Heierli of the Swiss Agency for Development and Cooperation in New Delhi, India; Kamla Chowdhry in New Delhi; Francis Steyn and his colleagues in South Africa; Anita Alvarez de Williams in Mexicali, Mexico; and the Cocopa community of El Mayor in the Colorado delta of northern Mexico.

Sometimes years go by before we recognize the influence certain people have had on our life's course. This was the case for me with Professor Cynthia Behrman, a scholar and teacher of great integrity and skill who taught history at Wittenberg University in Springfield, Ohio. Her confidence in and nurturing of my intellectual abilities came at an important moment in my life. I also owe a debt to Professor Thomas Gerrard, who not only inspired me to study geology but supported my interests in writing. As I returned with pleasure to the Wittenberg campus a few weeks ago to deliver the 1999 IBM-Endowed Lecture in the Sciences, I was reminded again of the value of the liberal arts tradition and of how it had broadened my world view.

While writing *Pillar of Sand*, I benefited beyond measure from the support, good humor, and votes of confidence that can only come from close friends and family. I thank Susan Embree-Davis, Harold and Clara Postel, Judith Vickers, Frederik van Bolhuis, Elaine Hanson, Linda Harrar, Andrea Fella, Joe Keyser, and Kevin Quinn. Very special thanks to Amy Vickers for inspiration in conceiving of this book and unparalleled support while writing it.

Finally, I would like to acknowledge Dorothy Anna Borcherding Postel. Although she died many years ago, when I was a young teenager, her sacrifices for me and pride in me have borne much fruit—not least of which is *Pillar of Sand*. With this book, I honor the life and memory of my mother.

Sandra Postel

I met a traveller from an antique land
Who said: "Two vast and trunkless legs of stone
Stand in the desert. Near them on the sand,
Half sunk, a shattered visage lies, whose frown,
And wrinkled lip, and sneer of cold command,
Tell that its sculptor well those passions read
Which yet survive, stamped on these lifeless things,
The hand that mocked them, and the heart that fed.
And on the pedestal these words appear:
'My name is Ozymandias, King of Kings:
Look upon my works, ye Mighty, and despair!'
Nothing besides remains. Round the decay
Of that colossal wreck, boundless and bare
The lone and level sands stretch far away."

Percy Bysshe Shelley
Ozymandias

PILLAR OF SAND

I

NEW LIGHT ON
AN OLD DEBATE

═══════════════

The sage's transformation of the World arises
from solving the problem of water.

Lao Tze

It is impossible to talk about the history of human civilization without talking about water. The story of settled agriculture, the growth of cities, and the rise of early empires is to no small degree a story of controlling water in order to make the land more prosperous and habitable. From ancient Sumeria and Babylonia to twentieth-century India and the United States, leaders through the ages have viewed large river engineering schemes as key to advancing their societies and bolstering political power. In 1997, when Egypt's President Hosni Mubarak unveiled a large new canal project to water his nation's western desert, he was following in the footsteps of a long line of Egyptian rulers who made the inauguration of dams and canals a celebration of national pride and progress. The famous relief of the mace head of the Scorpion King, dating back 5,000 years, depicts one of Egypt's predynastic rulers holding a hoe, ceremoniously cutting a ditch.

When it comes to water, nature has dealt a difficult hand. In many of the world's best places for agriculture—sunny, warm, and fertile—rainfall is too scarce or unreliable for steady crop production. Many of the world's rivers are tempestuous and erratic—running high when water is least needed, and low when it is needed most. Much water resides underground, hidden from view, and requires vast amounts of energy to bring to the surface. Earth may be a water planet, but that has not made the task of putting water to human use an easy one.

Settled agriculture began some 10,000 years ago in the northern Mesopotamian highlands, where enough rain fell to grow sufficient food. The story of controlling water for agriculture, however, begins several thousand years later, around 4000 B.C., when a band of adventurous farmers migrated south into the plains between the Tigris and Euphrates Rivers, the heart of what would later be called the Fertile Crescent, in present-day Iraq. In a place called Eridu, not far from the Persian Gulf, these pioneers began to adapt to a new set of conditions.

They expanded their diets with fish and waterfowl, which were abundant in the river and its surroundings. After the Euphrates River flooded in spring, they sowed seeds in the marshy floodplains, continuing the practice of crop cultivation their ancestors had carried out for millennia in the highlands. But there was one big difference: this place received much less rain. The new settlers watched their crops sprout and grow, but then wither from dryness before harvest time. Their remedy for this dilemma was simple, but it had profound and lasting effects. They dug a ditch and diverted some of the river's flow in order to water their crops during the dry season. In this way, the plains of Mesopotamia and the ingenuity of its early settlers gave rise to the practice of irrigation.

Irrigation transformed the land like no other activity up to that time. By artificially applying water to their fields, farmers found they could grow an extra crop. Areas that were too dry to support crops at all could be turned into productive fields. For the first time, large food surpluses appeared, freeing a portion of society to pursue other activities. The inventions and advances of this nonfarm class spanned metallurgy, weaving, ceramics, specialized crafts, writing, architecture, and mathematics. Societies became more stratified as the range of social activities widened and the need for centralized management grew. Populations increased in size and density, producing the first true cities. In short, irrigation unleashed a profound transformation in human development, and created a new foundation from which civilizations sprung and blossomed.

At the same time, irrigation brought with it new vulnerabilities. It made agriculture dependent upon a network of hydraulic infrastructure—including dams, canals, and levees—that was a natural target of enemies. The infrastructure also needed to be maintained, which required vast amounts of organized labor, some of it provided by slaves. Increased population pressure and competition for resources contributed to militarism and regional warfare.

Irrigation's transformation of the land also brought with it the continual threat of soil degradation—in particular, the buildup of salts. In dry climates, evaporation of water from the upper layers of soil can leave behind in the root zone a layer of salt that is damaging to crops. This problem crept up surreptitiously on the ancients and chipped away at the foundations of the advanced societies they had created. Overall, the irrigation base, upon which everything else rested, required constant vigilance; if it was neglected, a cascade of destabilizing effects could unfold.

The role of irrigation in the rise and demise of civilizations over the last 6,000 years is much more than a historical curiosity. On the cusp of a new millennium, human society is now as dependent on this ancient practice as ever. At the dawn of the modern irrigation age, in 1800, global irrigated area totaled just 8 million hectares, an area about the size of Austria; today, the irrigation base is 30 times larger, encompassing an area 2.5 times as large as Egypt. We now derive about 40 percent of our food from irrigated land. Many agricultural experts are counting on such lands to provide the bulk of the additional food that will be required over the next three decades. Yet as described in later chapters, there are myriad signs that our modern irrigation base is just as vulnerable as at any time in the past—and these problems have surfaced with unsettling speed.[1]

Groundwater is being pumped faster than nature is recharging it in many of the world's most important food-producing regions—including many parts of India, Pakistan, the north China plain, and the western United States. In many river basins, particularly in heavily populated Asia, there is simply little "undeveloped" water to tap. Worldwide, the amount of irrigated land per person has been declining for nearly two decades because of the rising economic, social, and environmental costs of large new water projects. One out of every five hectares of irrigated land is losing productivity because of spreading soil salinization. And as water becomes scarce, competition for it is increasing—between neighboring states and countries, between farms and cities, and between people and their environment.[2]

Water scarcity is now the single biggest threat to global food production. Just two decades ago, serious water problems were confined to manageable pockets of the world. Today, however, they exist on every continent and are spreading rapidly. More than a billion people now live in countries or regions where there is insufficient water to meet modest food and material needs per person. In many of these areas, populations are expected to expand greatly over the next few decades, raising the prospect of greatly increased food-import needs. But poverty levels raise doubts about the ability of these nations to import enough grain to fill their emerging food gaps. Even so, global food models to date largely ignore water constraints, and as a result they present an overly optimistic picture of future food availability.[3]

Concern about possible missing links in the global food

outlook brought an eminent group of scientists, researchers, and officials to the Airlie House retreat center in the Virginia countryside in the early spring of 1997. The participants converged from international research centers, universities, United Nations agencies, philanthropic foundations, corporations, and development organizations to focus on a simple question: What factors will most influence whether or not everyone has enough food to eat in 2025?[4]

The session was more than an academic exercise, because it came on the heels of more than half a decade of disarmingly slow growth in the yields of the world's major cereal grains—the staples of the human diet. Whereas grain yields rose an average of 2.1 percent a year between 1950 and 1990, that annual increase dropped to 1 percent between 1990 and 1998. Eight years do not constitute a long-term trend, but the falloff raises a red flag about future food supplies. Why is the slowdown occurring? Is it a temporary blip or the onset of a long downturn? What response is required?[5]

The debate about whether food production will keep pace with population growth has waxed and waned during the 200 years since an inquisitive English clergyman named Thomas Robert Malthus penned his famous essay on population. Malthus's proposition, simply put, was that because population grows exponentially while food production grows linearly, the former would eventually outstrip the latter. He postulated that increased death rates from hunger, disease, and famine would maintain the balance between human numbers and food supplies.[6]

At the time, Malthus did not foresee how migration, industrialization, and technological advances would combine to ward off such a dire situation in his homeland. But the essence of his argument stayed alive. In 1898, exactly 100 years after Malthus published his essay, Sir William Crookes delivered a speech to the British Association for the Advancement of Science entitled "The Wheat Problem." He warned that unless scientists found new ways to boost world grain yields, starvation would spread widely by the 1930s.[7]

Today, a century later still, the debate continues—but with two important differences. First, the race between population growth and technological advancement has greatly picked up speed. When Malthus wrote his first essay, world population was about 900 million. Farmers, scientists, and engineers had 160 years to figure out how to feed the next 2 billion people. The jump from 3 billion to 5 billion, however, took less than 30 years. We are already halfway through the next increase of 2 billion, to 7 billion, which we will hit around 2015. Even though the pace of annual population growth has slowed considerably—from a peak of 2.2 percent in 1964 to 1.4 percent in 1998—some 80 million people annually join humanity's ranks, roughly equivalent to adding another Germany each year.[8]

Second, until recent decades, additions to the world's food supply came from all three major food sources—rangeland, cropland, and fisheries. But two of these three already have hit or exceeded their natural limits. Overgrazing by livestock has caused as much as 20 percent of the world's pasture and range to lose productivity, which suggests that the global grass-eating livestock herd, now numbering about 3.3 billion, is unlikely to increase much, if at all. For the most part, future increases in meat production will come from feedlots rather than rangelands, adding to pressures on the grain supply.[9]

Likewise, overfishing has depleted natural fish stocks. The U.N. Food and Agriculture Organization (FAO) reports that harvests from 11 of the 15 most important fishing areas have either reached or exceeded their natural limits. On a per capita basis, the wild fish catch from marine and inland waters in 1997 was down nearly 8 percent from the 1988 peak. As with meat production, increases in fish supplies will need to come from fish farms, which, like feedlots, will require land and grain.[10]

With the spotlight turned on cropland, the picture becomes more complex and the crystal ball murkier. Most analysts agree that the worldwide cropland area will not expand much on a net basis. Each year, an estimated 10 million hectares are lost to erosion, other forms of degradation, or conversion to factories, houses, shopping malls, or other uses. FAO reports that total cropland area expanded an average of 1.6 million hectares per year between 1979 and 1994, but since losses are often not fully counted in official statistics, net cropland expansion could well be close to zero. Moreover, possibilities for opening up new cropland are mostly in areas where the long-term crop production potential is relatively low and the ecological costs—including forest loss and extinction of plant and animal species—are high, such as in parts of Brazil and central Africa.[11]

The principal remaining agricultural frontier is land productivity—coaxing more production from each parcel of cropland. However, researchers and farmers have already exploited a good bit of this frontier as well. Between 1950 and 1997, the area planted in grain expanded by only 17 percent even while total grain production rose by 190 percent. It was a spectacular 2.5-fold increase in grainland productivity over this period, supplemented with greater fish harvests and larger livestock herds, that allowed food production to keep up with population growth, keeping the Malthusian nightmare at bay.[12]

Today, the difference between the Malthusian pessimists and the cornucopian optimists comes down to little more than an assumption about grainland productivity over the next several decades—specifically, whether yields will grow at closer to the 1 percent rate of the 1990s or the 2 percent rate of the previous four decades. This difference may seem small, but it gets magnified over time. At an annual growth rate of 2 percent, food production would double in 35 years. Growing at 1 percent a year, however, it would take twice as long—70 years—to double.

The path food production follows will depend to no small degree on water. Crops cannot reach their maximum yield potential if they do not get sufficient moisture. As long as a crop receives ample sunlight and nutrients, its growth is directly linked to how much water it takes up through its roots and then releases through its leaves to the atmosphere, the process known as transpiration. How much water a crop consumes varies not only with the type of crop—wheat or rice, cotton or corn—but where it is grown. A rice plant growing in a hot dry place such as California or Egypt, for example, will need more water than one growing in a more humid location, such as the Mississippi delta or tropical Indonesia.

The global harvest of grains, oilseeds, fruits, vegetables, and other crops requires an enormous quantity of water. It takes about 1,000 tons of water, for example, to grow one ton of wheat. Worldwide, crops currently get about 70 percent of their water directly from rainfall and 30 percent from irrigation. Future increases in production, however, will rely much more heavily on irrigation because most of the grainland with abundant and reliable rainfall, such as the U.S. Corn Belt and Western Europe's major wheat areas, is already producing close to its maximum potential. Agricultural specialists expect production in these regions to increase somewhat as a result of efforts to breed or bioengineer plants that are more tolerant of drought, pests, or disease, but they do not foresee large leaps in yield in the near future. In drier areas, farmers will only purchase better seeds, apply fertilizers, and plant a second or third crop during the year if they feel fairly certain that they will have sufficient water to make those investments pay off. Achieving that certainty requires irrigation.[13]

All told, reaching the food production levels needed in 2025 could require up to 2,000 cubic kilometers of additional irrigation water—a volume equivalent to the annual flow of 24 Nile Rivers or 110 Colorado Rivers. As later chapters will show, supplying this much additional water will be difficult. The modern irrigation age—characterized by the engineering of whole river basins and mechanized control over vast quantities of water—

is running out of steam.

The best sites for dams are taken; numerous groundwater reservoirs already are overtapped. Existing dams and river diversions have wiped out vital habitat, decimating fish populations and pushing numerous aquatic species to the brink of extinction. Reservoirs are filling with silt. Fertile soils are slowly being poisoned by salt. Tensions are rising among nations that share common rivers, as they realize there is not enough water to satisfy all their demands. Bloated, inefficient bureaucracies have managed irrigation systems poorly, and have often allowed schemes to benefit rich and politically powerful farmers more than the poor. Large subsidies for irrigation have not only worsened government budget deficits, they have encouraged wasteful water practices that are flagrant anachronisms in today's world of scarcity.

In short, our irrigation base is showing numerous signs of vulnerability just as we are about to become even more dependent on it. Many of the same insidious threats that undermined ancient irrigation civilizations—including salt, silt, neglect of infrastructure, regional conflict, and unexpected climatic change—are rearing their ugly heads. Unless we transform irrigation again, and confront the consequences of large-scale water engineering, irrigation's environmental price will rise markedly, its productivity will deteriorate, and it will not expand food production to the degree needed for all to be fed.

This book makes the case that we need to double water productivity—get twice as much benefit from each liter of water we remove from rivers, lakes, and underground aquifers—if we are to have any hope of fulfilling the water requirements of 8 billion people and protecting the natural ecosystems on which economies and life itself depend. Meeting this challenge will involve making irrigation leaner and smarter—substituting knowledge and better management for water. It will involve spreading the whole spectrum of water-thrifty technologies that enable farmers to get more crop per drop. And it will require fixing a flagrant flaw of the modern irrigation age—the failure to provide technologies and methods that allow the smallest and poorest farmers to share in irrigation's benefits. Particularly in South Asia and sub-Saharan Africa, access to irrigation is a key to boosting food production and incomes for many of the 840 million people who are hungry and undernourished today.[14]

Perhaps the biggest challenge is bucking complacency. Food prices are at historically low levels, thanks in part to the increases in food output that irrigation helped generate. These prices, however, make it hard to justify new investments. As a result, the world's irrigation assets, representing a total capital value of $1.9 trillion, are dangerously prone to neglect. Moreover, in this age of space travel, instantaneous Internet communications, and life-prolonging medical advances, it seems pedestrian to worry about something as simple as having enough water to meet the world's food needs. But as Harvard anthropologist Timothy Weiskel reminded a U.S. Senate committee a decade ago, "There is no such thing as a post-agricultural society." To act as if there were is a recipe for societal collapse.[15]

Irrigation's historical record spans six millennia. The modern experience of the last two centuries is not only a young experiment, but one of uncertain outcome. The overriding lesson from history is that most irrigation-based civilizations fail. As we enter the third millennium A.D., the question is: Will ours be any different?

2

HISTORY SPEAKS

*Knowledge of the past helps to
anticipate the future.*

Thucydides

At least a half-dozen major irrigation-based civilizations arose between 2,000 and 6,000 years ago. Several of them, including the Sumerians, Babylonians, and Assyrians, thrived in the basin of the Tigris and Euphrates Rivers of present-day Iraq. The Egyptians formed the longest-lasting irrigation society, in the valley of the Nile. Distinct civilizations developed in the Indus River valley of present-day Pakistan and in the Yellow River basin of north-central China. Somewhat later, irrigation-based cultures arose in the western hemisphere as well. Central Mexico, coastal Peru, and the American Southwest each saw the rise and fall of an advanced society rooted in irrigated agriculture. The role of irrigation in shaping these societies, and the common threads of vulnerability and environmental deterioration these cultures exhibited, offer some lessons for our modern, globally integrated irrigation society today.

Breaking New Ground in Mesopotamia

The people who migrated out of the Mesopotamian highlands some 6,000 years ago and settled in the lowland plains of the Fertile Crescent came to be called Sumerians. Early on, their principal preoccupation was determining how to thrive in a place highly prone to both flooding and dry spells while making optimal use of the ample sunshine and warmth of their new climate.

The Tigris and Euphrates Rivers provided abundant water to hedge against drought, but their cycles of natural flooding did not coincide well with the cropping season. Peak flows tended to occur in April for the Tigris and in May for the Euphrates, just when the grain was ready for harvesting. Not only was irrigation unnecessary at this time, but high floods could wipe out their crops. To make matters worse, the flood-waters typically receded in June, just when the hot season was getting under way and extra soil moisture for new plantings was most needed. So the Sumerians faced the dual challenge of taming the floods and diverting water from the river for irrigation during the summer cropping season.

Exactly how they managed to do this remains obscure, but the growth and development of early Sumerian society attests to their success. With the emergence of a number of population centers and the production of sizable food surpluses, the Sumerians built the world's first urban society. By 3000 B.C., Sumer was dominated by eight cities, several of which had populations of 10,000–20,000 or more. In addition to Eridu, the southernmost Sumerian city and probably the earliest, there were Kish, Ur, Lagash, Umma, and Uruk. (See Figure 2–1.)[1]

Early on, priests had the task of managing food surpluses, and the temples served as storehouses for the collection and redistribution of grain. The greater the surplus, the larger the number of people who could pursue nonfarm activities—and

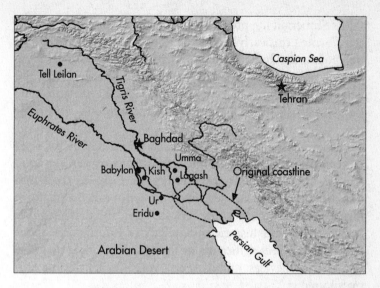

Figure 2–1. Ancient Mesopotamia

the more stratified the society became. Slaves constituted the bottom of the class hierarchy, followed by masses of peasant farmers, and then by craftspeople. Administrative, religious, and military elites formed the upper rungs.[2]

This pattern of social stratification and centralized control were traits not only of Sumer but of several other early irrigation civilizations as well, which led historian Karl Wittfogel to put forth his theory of the "hydraulic society." According to Wittfogel, large-scale river irrigation demanded mass labor that had to be "coordinated, disciplined, and led," and thus produced not only class divisions, but despotic forms of political control. Wittfogel viewed large-scale irrigation as the force behind predictable patterns of societal development. Although discounted in more recent times for its lack of historical precision, Wittfogel's theory drew attention to the relationships between large-scale irrigation and the evolution of social and political structures.[3]

As Sumerian life progressed, a host of inventions gradually transformed this early society and catapulted human civilization forward. The wheel was initially developed for making pottery, probably around 4500 B.C., and was later used to build wheeled vehicles drawn by domesticated animals. Around 3300 B.C., Sumerians made their greatest contribution—the invention of writing.[4]

This intellectual advance was likely stimulated by the need to keep track of all the temple transactions concerning food storage and distribution. By pressing sharp-tipped reeds into wet clay, the Sumerians developed the script called cuneiform that enabled them to record their activities, and that later gave archeologists and historians a wealth of information about their early society. Other major Sumerian advances included water-lifting devices, sailboats, and yokes for harnessing animals to plows.[5]

Sometime around 2300 B.C., a minister of the king of Kish, called Sargon of Akkad, took control of the independent cities and villages of Sumer, and established the first known human empire. Akkadian rule expanded to encompass 1,200 kilometers from the irrigated plains of southern Mesopotamia to the rainfed agricultural plains of northeastern Syria. Despite its considerable spread and wealth, the empire prospered for only 100 years. From the nature of its collapse, we begin to glimpse some of the inherent vulnerabilities of irrigation-based societies.

The Sargon lineage of rulers conquered the northern rainier plains in large part for the wheat, barley, and sheep the area could supply to their expanding empire. By this time, the population centers of these northern plains had grown from small agricultural communities, where some of the earliest settled farming had taken place, to sizable cities with structured economies. The area of a city called Tell Leilan, where much of the revealing archeological work has taken place, had expanded sixfold over 200 years.[6]

Excavations on the Tell Leilan Acropolis have turned up

evidence that a central administration collected, stored, and redistributed wheat, barley, and other foodstuffs. Impressions on jar sealings found at the site show renderings of banquet scenes from southern Mesopotamia, establishing that the northern rain-fed lands and southern irrigated lands had cultural contact. Both site excavations in the region and the cuneiform texts that have been found document that the imperial economy of the Akkadian empire was sufficiently thriving to support long-distance trade, the building of impressive palaces, and the construction of large agricultural projects. The scope and scale of Akkadian activities represented a degree of regional unification and economic integration that had not existed before.[7]

Then, abruptly, the empire collapsed. The best explanation for the sudden decline comes from a team of archeologists, soil scientists, and geologists who have found strong evidence of a shift to a drier climate starting at about 2200 B.C. and lasting for 300 years. Special analyses of the soil at Tell Leilan and a few other sites showed a lack of earthworm and insect activity, as well as large amounts of fine, wind-blown sands—telltale signs of intensified aridity and desiccation of the landscape. Archeologists had known for some time that Tell Leilan and other cities of the northern Mesopotamian plains had been abandoned for 300 years, but until these findings were reported in 1993, there was only speculation as to why. Exactly what caused the climate to shift remains unknown, but the climate piece made the rest of the puzzle of the Akkadian collapse fall into place.[8]

Prolonged drought made it impossible to grow grain without irrigation. As a result, tens of thousands of people streamed out of the northern rain-fed plains. These ecological refugees tried to join the cities and towns down the river valley in southern Mesopotamia just at the time the Euphrates's flow was diminishing because of the drop in precipitation in its watershed. The third dynasty of Ur, which succeeded the descendants of Sargon, recorded an influx of "barbarians" from the north and built the Repeller of the Amorites wall, a 175-kilometer barrier intended to block the advance of these unwanted migrants.[9]

Nevertheless, the population of the southern region doubled. To make matters worse, the reduced volume of the Euphrates's flow caused the river to drop more of its silt, which in turn must have led to the clogging of irrigation canals and water channels. Without the crops from the north that the empire had come to depend on, and with reduced irrigation capacities in the south, food supplies evidently fell considerably short of needs. The resulting famine and social disintegration brought the empire to its final downfall.[10]

We can only partially know the appropriate lessons to draw from the Akkadian demise. One overarching message is that a thriving, integrated, resourceful society can collapse abruptly from a cascade of ecological events. The irrigated agriculture that enabled this empire to expand well beyond anything previously known also became a source of vulnerability under the strains of population pressures and water shortages induced by abrupt climatic change.

The Scourge of Salt

While the drama of the rise and fall of the Akkadian empire was unfolding, a more insidious ecological threat emerged in the area—the buildup of salt in the soil, a process known as salinization. Salt's ability to poison the soil of irrigated lands has posed a constant risk in many irrigation-dependent societies throughout history, and remains a threat to our agricultural foundation today.

All river water and groundwater contains dissolved salts. During irrigation, plants take water up from the soil but leave most of the salts behind. In humid climates, rainfall percolating through the soil pushes the salts out of the root zone. But in drier climates, farmers must apply extra irrigation water to do this job. This additional water can lead to even greater problems, especially in low-lying river valleys, where much of the world's irrigation takes place. As more and more water seeps through the soil to the groundwater below, the water table rises. As it nears the surface, some of the water evaporates, leaving the salts behind. If the problem is not corrected, the buildup of salt poisons the land, rendering it toxic to crops.

During the 1950s, the government of Iraq became interested in improving its understanding of the historical experience with salinization, a problem that was looming large in its agricultural development plans. In 1957, in partnership with the Oriental Institute of the University of Chicago, the government launched a major archaeological project to study the early irrigation societies of the southern Mesopotamian plains. Drawing on both ancient texts and field excavations, the team pieced together a picture of just how serious a problem salt was to the ancients, and—by extension—how grave a threat it is today.[11]

The researchers uncovered evidence of three major episodes of salinization in ancient Mesopotamia. The earliest and most serious one affected what is now southern Iraq from 2400 B.C. until at least 1700 B.C.—a period that encompasses the rise and fall of the Akkadian empire and the permanent passing of power from the land of Sumer. While archeologists and Near East specialists do not agree on the full extent of salinization's role in ancient Mesopotamian history, it was a destabilizing force, if not a proximate cause of societal decline.

Early Sumerian farmers practiced simple flood and furrow irrigation, the methods still used on some 90 percent of irrigated land today. The Euphrates was easy to tap, because the river bed lay above the level of the surrounding plain, a common situation with rivers that carry and deposit a great deal of silt in their channels. Irrigators simply had to breach the levee alongside the river and water would flow into the main diversion canal. This canal would then supply water to smaller feeder-canals, which in turn delivered water to the irrigation ditches. Using a hoe or other simple tool, farmers would break open a gap in the wall of the irrigation ditch, letting water flow across their plots in a shallow flood or down parallel furrows. When the irrigation was completed, they would plug the opening in the ditch wall with mud, allowing the irrigation water to flow to the next plot, and then on to the next. The records of the rulers of Sumer and Akkad contain numerous references to the building of canals, the heart of this type of valley irrigation.[12]

The irrigators fallowed their plots every other year. This rest period gave their fields some time to regain fertility and for deep-rooted weeds to take hold and pull excess water out of the soil before the next planting. In this way, the rise in the water table and consequent salt buildup were kept in check. But these problems could not be eliminated altogether.

In addition, a long-standing dispute between two neighboring Sumerian cities probably worsened the salt problem. For generations, Girsu and Umma had fought over a fertile parcel of land near their border. Girsu's ruler, Entemenak, became frustrated with the obstructionist tactics of his upstream rival and decided to seek an alternate water source to irrigate the disputed territory. He built a large canal from the Tigris River, which brought copious quantities of additional water to this section of the lower Euphrates basin. The excess irrigation and seepage that went along with this abundant new supply apparently hastened the process of destructive soil salinization.[13]

One of the key pieces of evidence that salt buildup reached damaging levels sometime after 2400 B.C. is the change in the region's crop mix. Wheat was the preferred cereal for eating, but it is less tolerant of salt than barley is. Grain impressions found in pottery from southern Iraqi sites dating back to 3500 B.C. suggest that about equal amounts of wheat and barley were grown at that time. Just over 1,000 years later, wheat apparently accounted for little more than one sixth of the harvest. By 2100 B.C., wheat's share had dropped to less than 2 percent in the Girsu area, and by 1700 B.C. it was no longer cultivated at all. Moreover, at the same time the crop mix was shifting, yields of barley were declining—another sign that salt had poisoned the land.[14]

The southern portion of the Fertile Crescent never fully recovered from the decline that accompanied the buildup of salt. With the rise of Babylon in the eighteenth century B.C., political power shifted northward and never returned to the south. The former cities of Sumer, the earliest urban societies, were reduced to small villages or left in ruins. Archeologist Leonard Woolley, who wrote in the 1930s about his excavations of the city of Ur, struggled to make sense of the stark contrast between the thriving ancient civilization he helped unearth and the desolate wasteland he encountered in twentieth-century southern Iraq: "Why, if Ur was an empire's capital, if Sumer was once a vast granary, has the population dwindled to nothing, the very soil lost its virtue?"[15]

Decades of more digging and research would be required before the puzzle began to come together, and no doubt archeologists will continue to fill out what remains an incomplete picture. Thorkild Jacobsen and Robert M. Adams, both with the Oriental Institute when the Iraqi research program was carried out, concluded in a 1958 article in *Science* that there is probably "no historical event of this magnitude for which a single explanation is adequate, but that growing soil salinity played an important part in the breakup of Sumerian civilization seems beyond question."[16]

There is a Sumerian myth that offers a fitting epitaph to this historical account. In the tale, the goddess of death greatly envies her sister, the goddess of love and procreation. She seeks revenge by causing salty water to rise from below the earth to poison the life-giving soil.[17]

From Babylon to Ninevah

After the first serious episode of salinization in the Mesopotamian plains had pushed the demographic and political center northward, Babylon rose to become one of the greatest empires of antiquity. The powerful King Hammurabi, who ruled in 1792–50 B.C., united the warring states of the valley and established a historic code of law that was extremely advanced for its time.

The famous Code of Hammurabi contained 285 laws, and several of them dealt with irrigation. Landholders, for example, were required to keep their sections of irrigation ditches in good repair. If they failed to do so and their neighbor's fields were flooded as a result, they had to pay compensation for the harm done. In the code's prologue, Hammurabi reveals his self-perception as a great provider, calling himself "the exalted prince...who supplied water in abundance to [Uruk's] inhabitants;...who stored up grain for the mighty Urash;...who helped his people in time of need."[18]

Indeed, the irrigation works built under Hammurabi's rule had much to do with the greatness Babylon achieved. The king had a large canal built between the old Sumer city of Kish, just south of Babylon, all the way to the Persian Gulf—greatly expanding the empire's irrigated area and affording flood protection to the southern cities. With the taxes he levied on his subjects, Hammurabi built a host of palaces and temples, constructed a bridge spanning the Euphrates so that the city could spread out on both sides, and encouraged a vibrant shipping trade up and down the river. Under Hammurabi's centralized power and codified rule of law, Babylon became one of the richest cities the world had ever seen.

After Hammurabi's dynasty ended with an invasion by the Cassites from nearby mountains, several hundred years passed before the next major empire advanced irrigated agriculture further. The Assyrians' political center was located still further north, on the Tigris River, and their irrigation laws, building on those of Hammurabi, implied a strict social contract. These laws set out both rights and responsibilities of landowners who received water from a common source. They mandated cooperation in keeping irrigation canals free of silt, protecting the supply, and making sure those farthest from the water source received their fair share of water—an equity issue in many irrigation schemes today. Landowners were required to offer labor in proportion to the size of their property. A special court was set up to enforce all these obligations.[19]

Like the Babylonians before them, the Assyrians took great pride in building irrigation canals and boosting agricultural production. Since their climate was much wetter than that of the Babylonians and Sumerians, they had previously had little experience with perennial irrigation, but they happily acquired it. Queen Sammu-Ramat, who ruled Assyria briefly during the late ninth century B.C., had inscribed on her tomb: "I constrained the mighty river to flow according to my will and led its water to fertilize lands that had before been barren and without inhabitants." About 120 years later, in 691 B.C., King Sennacherib built what historians of technology view as one of the most impressive works of hydraulic engineering until Roman times—an 80-kilometer canal, paved with masonry, designed to bring additional water to his capital, Nineveh. Just outside the city, a dam diverted a portion of the aqueduct's water into side-channels for irrigating beautiful orchards and gardens.[20]

The Assyrian rulers built an extensive network of canals, along with small reservoirs for storing water that flowed out of the surrounding hills. In these uplands of what is now northern Iraq, rainfall sufficed to grow cereal grains and vegetables, but the Assyrians developed perennial irrigation for more intensive crop cultivation in heavily populated areas, as well as for their orchards and gardens. A riverine scene found at Sennacherib's palace at Nineveh suggests that the Assyrians made good use of the *shaduf*, an ancient water-lifting device that dates back at least to the third millennium B.C.[21]

A typical *shaduf* consists of a horizontal beam fixed across two pillars over which a long slender pole pivots. At one end of the pole a bucket is suspended, and at the other end a large mound of clay acts as a counterweight. Standing next to the river or canal, the operator fills the bucket with water, and then allows the counterweight to lift it back up. Just as the bucket nears the peak of the upward swing, about waist level, the operator tips the bucket into an irrigation ditch, which carries the water to nearby fields. With a *shaduf*, a person could lift 600 gallons of water a day—much more than any nonmechanical method—which explains why it became such a mainstay of irrigation in ancient Mesopotamia, Egypt, and elsewhere.[22]

In part because of irrigation successes, the population of Assyria increased greatly during the seventh century B.C., although exact numbers are not known. Assyrian farmers cultivated a host of plants for eating and medicinal purposes, including turnip, leek, garlic, onion, mustard, radish, lettuce, cucumber, fennel, coriander, mint, rosemary, and ginger. The first mention of cotton in ancient times dates to this empire.[23]

While the Assyrians advanced the technology and practice of irrigated agriculture, they also developed a penchant for punishing defeated enemies by destroying their irrigation works. Sargon II, Sennacherib's father, wrecked the irrigation system of an enemy city by damming the main canal and purposely causing a flood. After razing Babylon, Sennacherib dumped the debris from temples and palaces into the irrigation canals and then flooded the whole city: "I completely blotted it out with water-floods and made it like a meadow." Along with their advances, the Assyrians thus provided some of the earlier examples of the use of water as a weapon of war.[24]

The Final Fall

Mesopotamian society reached its zenith between the third and seventh centuries A.D. under the rule of the Sassanians, a powerful family from the Iranian highlands who conquered the region around A.D. 220. Their rule extended not only throughout the vast Mesopotamian alluvial plain of present-day Iraq, but also into southwestern Iran. They planned and invested in a series of bold water schemes that expanded irrigation to nearly all the arable land in the region. Archeologists estimate that the area under cultivation by late Sassanian times totaled some 50,000 square kilometers, 40 percent more than Iraq's total irrigated area today. The population at that time is not known, but based on the labor that the irrigation systems would have required, historians place it at roughly 5 million.[25]

In the centuries that followed, after the Arab conquest in A.D. 639, the population declined dramatically and the tale of the great Mesopotamian societies essentially came to an end. This last piece of the story, which concludes about 500 years ago, best illuminates the inherent vulnerabilities of irrigation-based societies.

First, silt became a major problem. Both the Tigris and the Euphrates carry enormous loads of silt. At flood stage, they can transport as much as 3 million tons of suspended sediment in a single day. As the rivers meander across their plains, which slope very gradually toward the Persian Gulf, they lose speed, which causes them to drop their sediment loads. Over time, these deposits elevated the river beds and banks above the surrounding plain. This made the rivers highly prone not only to flooding, but also to changing course, which historically they have done many times. Some of the silt also got deposited in the irrigation canals and had to be removed periodically to avoid clogging the system. This work was extremely labor-intensive: during his reign over the Mesopotamian plains in the fourth century B.C., Alexander the Great put 10,000 people to work for three months cleaning out and repairing one diversion canal on the west side of the Euphrates.[26]

In addition to siltation, salt buildup again became a major problem. A vast network of canals crisscrossed the Mesopotamian plains, including several designed to transfer water between the Euphrates and the Tigris. These canals were oriented perpendicular to the direction of natural drainage, which formed artificial basins that were highly prone to waterlogging and salinization. By the seventh century A.D., salt buildup had reached damaging levels in parts of the plain. Records show that 15,000 slaves were forced to work in the southern region peeling off the upper sterile layers of soil to get to more-fertile layers below. Because of their intolerable working conditions, the slaves revolted periodically during the seventh and eighth centuries, culminating in a 14-year uprising from A.D. 869–883. This revolt, and the famines that occurred along with it, took more than a million lives.[27]

At about the same time, epidemic disease further decimated the work force. Most of the irrigation maintenance was done not by slaves but by the *corvée*—an army of unpaid laborers from the peasantry who had to serve at the state's request. If war or disease depleted the ranks of this hydraulic army, irrigation canals could clog with silt and flood damage could go unrepaired. Ten years passed, for example, before a breach in one of the canals connecting the Euphrates with the Tigris was fixed. Such breakdowns could leave large areas without needed irrigation water, reducing crop production. Coupled with declining yields caused by salt buildup in the soil, the problems of siltation and neglected maintenance put the food system at risk.[28]

Whether salinization and environmental deterioration alone would have eventually caused Mesopotamia's final downfall will never be known. What is certain, however, is that the inherent environmental instability of the irrigated agriculture these societies depended on made them highly vulnerable to even small disturbances. As scholar Peter Christensen concludes in *The Decline of Iranshahr*, ultimately the system could not withstand "the clash between the environmental and technological limitations of agricultural production and the political demands for surplus."[29]

Agricultural tax assessments from archives in Baghdad record the final fall of Mesopotamian society. The tax receipts do not correlate directly with the volume of agricultural output over time, but there is enough correspondence between them to document the decline. For a period of 700 to 900 years, tax revenues were relatively stable at about 100 million dirhams. By the beginning of the tenth century A.D., however, receipts had plummeted to 30–40 million dirhams—a drop that must have led to further neglect of irrigation works and land reclamation.[30]

This revenue decline is particularly striking in light of the diverse mix of crops that were being grown at the time, including plums, pears, melons, pomegranates, olives, and citrus. The cultivation of sugarcane, with its long growing season, shows that farmers had the capacity for year-round irrigation. Rice, a thirsty summer crop, was a dietary staple. Taken together, these crops attest to the advanced cultivation methods and irrigation systems that were in use. But these same intensive practices also greatly strained the environment. There is evidence of another major episode of salinization east of Baghdad sometime after A.D. 1200.[31]

Ultimately, the agricultural base could not hold up amid the broader social and political changes that were occurring. By the fourteenth century, after the sack of Baghdad by the Mongols, agricultural tax assessments totaled just 13 million dirhams, an 87-percent drop from their peak.[32]

By the sixteenth century, the Fertile Crescent of Mesopotamia, from which human civilization had sprung and reached unprecedented heights, was little more than a salty wasteland. These societies had given the world writing, mathematics, magnificent palaces, and unparalleled feats of engineering. But they had not created a system of agriculture that could sustain their people.

Wellsprings of the Indus and Yellow Rivers

Compared with their Mesopotamian counterparts, relatively little is known about the history of the great irrigation societies of the Indus valley and China's Yellow River basin. But they were also cradles of human civilization in their corners of the world.

Ancient Chinese efforts to control and tap the water resources of the north China plain appear to have begun about 4,000 years ago. According to Chinese legend, around that time the Yellow River broke through its banks and flooded the surrounding plains. A man called Yu, subsequently surnamed the Great, organized workers to build dikes along the river bank and to remove the silt clogging the river channels, thereby bringing the river temporarily under control. He gave the reclaimed land to farmers for crop cultivation. Today, a statue of Yu the Great stands in the lower basin in recognition of the prosperity this ancient leader brought to the people of the Yellow River valley.[33]

Spanning some 300,000 square kilometers, the north China plain is situated east of the highly erodible Loess Plateau and was built up over the millennia by the rich deposits of silt brought down from the highlands. Around 1750 B.C., the entire plain came under a single ruler and China's first urban, stratified society emerged. As in Mesopotamia, there was strong centralized control of labor, as well as centralized mechanisms for storing and redistributing grain.[34]

Diversions from the Yellow River to irrigate paddy rice go back to the early Han Dynasty of 200 B.C.–A.D. 200. The Yellow, however, proved a tough river to tame. It emerges from the Loess Plateau carrying some 1.6 billion tons of silt each year, which makes the river extremely prone to channel alterations and flooding. The Yellow breached its dikes more than 1,500 times over the 2,000 years prior to 1950. The many devastating floods that resulted earned it the nickname China's Sorrow.[35]

Because the heavy silt deposits raise the river above the surrounding plain, the river frequently changes course—on average about once a century. These unpredictable shifts have affected the stability of agriculture and human settlement on about three quarters of the north China plain. They have also forced the Chinese to spend enormous sums to continually raise and reinforce dikes to hold back the floodwaters. It is a testament to the ingenuity and hard work of the ancient Chinese that they could create a brilliant civilization in the valley of such an uncooperative river. Still, by about A.D. 600, the Yangtze River valley had replaced the north China plain as the economic center and grain basket of China.[36]

Even less is known about the early Indus civilization, although—like the ancient Mesopotamian and Chinese cultures—it too depended on perennial irrigation. The Indus valley, which spans part of present-day Pakistan and India, was likely settled by migrant farmers moving east out of southwest Asia around 3500 B.C., and development there was probably influenced by that of Mesopotamia. This culture, too, evolved into a stratified society once sufficient food surpluses were generated, and by about 2300 B.C. a powerful civilization existed.

Early Indus society appears to have been culturally more uniform. Its two major cities—Harappa and Mohenjo-Daro, located some 640 kilometers apart—were built according to similar plans. Historians do not know whether the central authority of this civilization was religious, as in Mesopotamia, or secular, as in China, but it was authoritarian and able to mobilize vast numbers of laborers. As with its early counterparts, the Indus rulers maintained central granaries for storing and redistributing food.[37]

The complex, hierarchical society that developed in the

Indus valley lasted less than 500 years. Because the script used for recordkeeping has not been deciphered, it is impossible to know exactly what brought this civilization to collapse. Some of the same forces at work in Mesopotamia were probably also operating in the Indus valley. Salinization was almost certainly a problem, as it still is today. The environment likely also suffered from the choice of construction materials for temples, palaces, and other city structures. Like the Mesopotamians, the Indus valley inhabitants used mud bricks, but rather than drying them in the sun, they baked them in wood-fired ovens. Many trees must have been cut in the surrounding watersheds, which in turn would have increased both flooding and siltation in the valley below.[38]

Although conquest by outside invaders may have been the immediate cause of the Indus collapse, instability caused by soil salinization, siltation, flooding, and possibly climatic change may have weakened the society from within and initiated a process of decline. British geographer Malcolm Newson points out that the size of the Indus civilization, with its two major cities located hundreds of kilometers apart, may have worked against it as well. It is possible, he notes, that "management failed to respond in a coordinated way to environmental change, however caused."[39]

Staying Power in Egypt's Nile Valley

In striking contrast to the early Indus civilization and those of Sumer, Akkad, Babylonia, and Assyria in Mesopotamia, the great Egyptian civilization in the Nile River valley has sustained itself for some 5,000 years without interruption. It lasted through warfare and conquest by the Persians, Greeks, Romans, Arabs, and Turks, as well as through pandemic disease that devastated its population. Yet its agricultural foundation remained intact. Only in more recent times has the sustainability of Egyptian agriculture come into question. In response to a 20-

fold increase in its population over the last two centuries—
from 3 million in the early 1800s to 66 million today—Egypt
replaced its time-tested agriculture based on the Nile's natural
flow rhythms with more intensified irrigation and flood management that required complete control of the river.[40]

Compared with the flashy floods of the Tigris and the
Euphrates, the historic Nile flood was much more benign, predictable, and timely. As is the case today, most of its flow orig-
inated from monsoon-type rains in the Ethiopian highlands.
The remainder came from the upper watershed of the White
Nile around Lake Victoria. With almost calendrical precision,
the river began to rise in southern Egypt in early July, and it
reached flood stage in the vicinity of Aswan by mid-August.
The flood then surged northward, getting to the northern end
of the valley about four to six weeks later.

At its peak, the flood would cover the entire floodplain to a
depth of 1.5 meters. The waters would begin to recede in the
south by early October, and by late November most of the valley was drained dry. Egyptian farmers then had before them
well-watered fields that had been naturally fertilized by the rich
silt carried down from Ethiopia's highlands and deposited on
the floodplain as the waters spread over it. They planted wheat
and other crops just as the mild winter was beginning, and harvested them in mid-April to early May. By this time, the river's
flow had diminished, sustained only by the more constant flow
of the White Nile; the floodplain was completely dry. Then,
magically to the ancients, the cycle started all over again. Even
into modern times, every June 17th Egyptians celebrated the
"'Night of the Drop,' when the celestial tear fell and caused the
Nile to rise."[41]

The Egyptians practiced a form of water management called
basin irrigation, a productive adaptation to the natural rise and
fall of the river. They constructed a network of earthen banks,
some parallel to the river and some perpendicular to it, that
formed basins of various sizes. Regulated sluices would direct

floodwater into a basin, where it would sit for a month or so until the soil was saturated. Then the remaining water would be drained off to a basin down-gradient or to a nearby canal, and the farmers of the drained plot would plant their crops.[42]

The earliest evidence of water control in ancient Egypt is the famous historical relief of the mace head of the Scorpion King, which dates to around 3100 B.C. It depicts one of the last predynastic kings, holding a hoe and ceremoniously cutting a ditch in a grid network. Besides attesting to the importance of these waterworks and the great ceremony attached to them, this picture confirms that Egyptians began practicing some form of water management for agriculture about 5,000 years ago.[43]

Egyptian irrigators did not experience many of the vexing problems that plagued the irrigation societies of Mesopotamia. The single season of planting did not overly deplete the soil, and fertility was naturally restored each year by the return of the silt-laden floodwaters. In some basins, farmers planted grains and nitrogen-fixing legumes in alternate years, which helped maintain the soil's productivity. Fallowing land every other year, which was essential in Mesopotamia, was thus unnecessary in the Nile valley.[44]

Neither was salinization a problem. The summer water table remained at least 3–4 meters below the surface in most basins, and the month or so of inundation prior to planting pushed whatever salts had accumulated in the upper soil layers down below the root zone. With salt buildup naturally checked and fertility constantly restored, Egyptian agriculturalists enjoyed not only a productive system, but a sustainable one.

For nearly 1,500 years Egyptian farmers cultivated about 800,000 hectares under this system of basin irrigation. The *shaduf*, the water-lifting device already in use in Mesopotamia, appeared in upper Egypt sometime after 1500 B.C. This technology enabled farmers to irrigate crops near the river banks and canals during the dry summer. This would have allowed the cultivated area to expand by 10–15 percent. A similar increase might have been afforded by the waterwheel, introduced sometime after 325 B.C. So by the time Egypt had become a breadbasket for the Roman Empire, some 1 million hectares of land were effectively under cultivation in the course of a year.[45]

The blessings of the Nile were many, but they did not come without some costs. A low flood could lead to famine, and too high a flood could destroy dikes and other irrigation works. Even a 2-meter drop in the river's flood level could leave as much as a third of the floodplain unwatered.[46]

The well-known biblical account of Joseph and the Pharaoh's dream is a reasonable reflection of the threat of famine that Egyptians periodically faced. Asked to interpret his ruler's dream, Joseph foretells seven years of abundant harvests followed by seven years of shortage, and advises the Pharaoh to begin storing massive quantities of grain to avert famine. During a period of disappointing floods between the reigns of Ramses III and Ramses VII in the twelfth century B.C., food shortages caused the price of wheat to rise markedly. Prices stabilized at a high level until the reign of Ramses X, and then fell rapidly as shortages eased by the end of the Ramessid Dynasty, about 1070 B.C.[47]

Because of the link between the Nile's flow level and Egyptian well-being, early on the ancient Egyptians developed a system for measuring the height of the Nile in various parts of the country. This monitoring allowed them to compare daily river levels with years past and to predict with some accuracy the coming year's high water mark. At least 20 "nilometers" were spaced along the river, and the maximum level of each year's flood was recorded in the palace and temple archives.[48]

In combination, the reliability of the Nile flood and the unpredictability of its magnitude rooted ancient Egyptians deeply in nature and fostered respect for order and stability. Rulers were viewed as interveners with the gods to help ensure prosperity. Father of all gods was the god of the Nile—Hapi—

who although male was portrayed with breasts to show his capacity to nurture.[49]

The Egyptians worshipped Hapi not only in temples, but through hymns:

> Praise to you, O Nile, that issues from the Earth, and comes to nourish Egypt...

> If his flood is low, breath fails, and all people are impoverished; the offerings to the gods are diminished, and millions of people perish. The whole land is in terror...

> When he rises, the land is in exultation and everybody is in joy...

> He fills the storehouses, and makes wide the granaries; he gives things to the poor.[50]

In contrast to the Mesopotamian civilizations, early Egyptian society did not centrally manage state irrigation works. Basin irrigation was carried out on a local rather than a national scale. Despite the existence of many civil and criminal codes in ancient Egypt, no evidence exists of written water law. Apparently, water management was neither complex nor contentious, and oral traditions of common law withstood the test of a considerable amount of time.

Although difficult to prove, the local nature of water management, in which decisionmaking and responsibility lay close to the farmers, was probably a key institutional factor in the overall sustainability of Egyptian basin irrigation. The many political disruptions at the state level, which included numerous conquests, did not greatly affect the system's operation or maintenance. While both slaves and *corvée* labor were used, the system's construction and maintenance did not require the vast numbers of laborers that Mesopotamia's irrigation networks demanded. The waves of plague and warfare that periodically decimated Egypt's population did not result in the irrigation base falling into serious disrepair, as occurred in Mesopotamia.

Local temples appear to have played an important role in redistributing grain supplies to help cope with the periodic famines. From very early times, boats plied the Nile and were used to transport grain from one district to another. The surplus from several districts might have been stored in a central granary and shared to secure food supplies for the whole region. Fekri Hassan, a professor in the department of Egyptology at the University of London, speculates that the emergence of kingship in Egypt was linked to the need for larger coordination in collecting grain and providing relief supplies to districts experiencing crop failure.[51]

The central government imposed a tax on the peasant farmers of about 10–20 percent of their harvest, but the basic administration of the agricultural system remained local. As Hassan observes, "Egypt probably survived for so long because production did not depend on a centralized state. The collapse of government or the turnover of dynasties did little to undermine irrigation and agricultural production on the local level."[52]

Overall, Egypt's system of basin irrigation proved inherently more stable from an ecological, political, social, and institutional perspective than that of any other major irrigation-based society in human history. Fundamentally, the system was an enhancement of the natural hydrological patterns of the Nile River, not a wholesale transformation of them. Although it was not able to guard against large losses of human life from famine when the Nile flood failed, the system sustained an advanced civilization through numerous political upheavals and other destabilizing events over some 5,000 years. No other place on Earth has been in continuous cultivation for so long.

From Tehuacán to Snaketown: Irrigation Societies
of the Early Americas

The Mesopotamian, Nile, Indus, and Yellow River civilizations stand out as the great ancient irrigation societies. Early irrigation's story is geographically incomplete, however, without a look at the societies that emerged in the western hemisphere. Though later in time and smaller in size, the irrigation civilizations that developed in coastal Peru, central Mexico, and southwestern North America shaped cultural developments in the Americas much as their better-known counterparts did in the Near East and the Orient.

If there is a geographic cradle of civilization in the western hemisphere, it is Mesoamerica, an area that encompassed what is today southeast Mexico, Guatemala, Belize, and parts of Honduras and El Salvador. Settled villages did not evolve in this part of the world until about 2000 B.C., when the productivity of domesticated corn reached a level that could support stable communities. The role of irrigation rose to the fore in the Tehuacán Valley, a dry highland region about 180 kilometers southeast of present-day Mexico City. Beginning around 300 B.C., the use of canals for irrigation "virtually exploded" throughout the basin, and, according to regional expert William Dolittle, developed into "a truly remarkable ancient canal irrigation system."[53]

To the south, in coastal Peru, the ancient Chimú Empire evolved into an advanced economy and complex society from an agricultural base that was entirely dependent on irrigation. Starting around 1000 B.C. and spanning nearly 2,500 years, it grew from small, priest-dominated communities based on irrigated corn agriculture to a gradually more sophisticated, secular, and urban empire—an evolution that in many ways paralleled that of ancient Mesopotamia.

Along the warm Peruvian coast, an irrigated crop could be raised in just four months. The Chimú rulers put the surplus time and labor to use in the construction of numerous pyramids and temples and in the development of fine textiles, ceramics, and handicrafts. They established a centralized bureaucracy to manage the large irrigation systems and to control the distribution of water. ChanChan, the empire's capital, may have reached a population of 50,000, and from it the rulers governed more than a dozen coastal irrigated valleys. At its peak, the whole empire probably numbered more than 500,000 people. Conquered by the Incas around A.D. 1470, this long-lasting ancient Peruvian empire ranks among the landmark cultures in human social evolution, and irrigation underpinned much of its success.[54]

Last but not least, the Hohokam round out this look at the world's early irrigation civilizations. Their culture thrived for more than 1,000 years in the basins of the Gila and Salt Rivers in what is now south-central Arizona—and then abruptly disappeared around A.D. 1400.

Archeologists studied the culture only sporadically until the 1970s, but then began a feverish round of exploration when massive migration into Phoenix and Tucson threatened to obliterate much that remained of Hohokam communities. Ironically, one of the largest modern intrusions into Hohokam territory was the Central Arizona Project, a vast canal system built to divert Colorado River water hundreds of kilometers to Arizona's thirsty and burgeoning metropolises. Unfortunately, despite the flurry of archeological activity over the last two decades, a complete rendering of the Hohokam story will never be possible, because much of their world lies buried under golf courses, cotton fields, and urban sprawl.[55]

Less well known than their Anasazi neighbors at Chaco Canyon and Mesa Verde, the Hohokam successfully farmed the lower Sonoran desert by tapping the region's rivers and tributaries to feed vast canal networks. Archeologists have documented more than 500 kilometers of main canals, with at least 300 kilometers of them operating contemporaneously in a linked network. Further away from the rivers, the Hohokam practiced variations of floodwater farming to raise corn and other crops. Some farmers cultivated alluvial fans—areas where runoff and sediment emerge from a higher plateau. This diversity of agricultural practices may have enhanced the society's overall stability, and may help explain its 1,000-year duration.[56]

The Hohokam's hallmark, however, was irrigation. As with the other ancient irrigation societies that preceded them, irrigation fostered cooperation among communities sharing a common canal. It generated food surpluses that allowed them to devote time and resources to arts, crafts, and handiworks. From a settlement called Snaketown, southeast of Phoenix, archeologists have unearthed beads, bracelets, etched and painted shell ornaments, beautifully painted ceramics, and various styles of decorated pottery. Less urban than most of their historical counterparts, the Hohokam lived in villages evenly spaced along the major river systems, which was beneficial for the labor-intensive and coordinated work of canal maintenance.[57]

At its zenith, Hohokam trade and interaction reached north to Chaco Canyon and south to Mesoamerica. Its population peaked somewhere between 50,000 and 200,000, and its territory spanned some 115,000 square kilometers, an area about the size of Guatemala. What made this thriving culture fade away? No one knows for sure. But many explanations include environmental change—in particular, a series of droughts and floods.[58]

Toward the end of the eleventh century, rainfall decreased on the Colorado Plateau, the source of runoff for regional rivers and the Hohokam irrigation canals. Many changes in cultural and settlement patterns date to this time, including the abandonment of Snaketown and a number of other ancestral sites. Arts and crafts industries diminished, and some key architectural features disappeared. Subsequently, water conditions improved, and both population and irrigation expanded, until another drought struck around A.D. 1275.[59]

On top of the destabilizing effects of drought, a period of floods began around 1325. More rainfall in the uplands would appear to be helpful to a desert agricultural society, but not if it comes at the wrong time or too rapidly. For the Hohokam, heavy rains could lead to flood surges that wiped out irrigation canals and damaged crops. Archeologists have found evidence of canal breaches and interruptions in canal use within large irrigation systems supplied by both the Gila and Salt Rivers. Serious flooding disasters may have occurred as often as every five years, which would have reduced crop production capabilities, diverted labor from other activities, and generally strained the Hohokam economy. One of the main canals near Snaketown was apparently rebuilt 10 times.[60]

An environmental link to the Hohokam's demise may never be proved. Archeologists have identified other possible causes, including shifting trade patterns and changing regional alliances. But water—too little, too much, or coming at the wrong time—seems to have played an important part.

As we enter the greenhouse century of climate change, with its greater likelihood of extreme floods and droughts, the Hohokam tale may be prescient. When they were in the midst of their 1,000-year run of an advanced, irrigation-based culture, the Hohokam probably had no idea how vulnerable their society actually was.

3

IRRIGATION'S MODERN ERA

God and Manifest Destiny spoke with one voice
urging us to "conquer" or "win" the West.

Wallace Stegner

Modern irrigation owes a debt to the Italian Renaissance artist Leonardo da Vinci, who was captivated by water. Not only did he create beautiful renderings of water in motion, as in *Deluge*, but his intellectual curiosity brought him new insights into water's movement across the landscape. Da Vinci's sixteenth-century studies of the Arno watershed in northern Italy led him to better define the relationship between a river's catchment and its flow, helping to establish the fundamentals of modern hydrology and river basin management.[1]

It took several hundred more years and much trial and error, however, before researchers worked out the principles of hydraulics and engineers developed mechanized water control technologies that together transformed irrigation from an art to a science. In 1800, just before the dawn of the modern irrigation age, world irrigated area totaled 8 million hectares, an area about the size of Austria. From this small starting base, it climbed fivefold during the nineteenth century. Much of irrigation's scientific and technical foundation was built during that century's latter half, and these advances made possible large new schemes, particularly in Asia. From a total of 40 million hectares in 1900, the global irrigation base grew to some 100 million hectares by 1950, a 2.5-fold increase. (See Figure 3–1.) By 1995, world irrigated area had grown two-and-a-half-fold again, to just over 255 million hectares—an area more than two-and-a-half times the size of Egypt.[2]

Today, four countries—India, China, the United States, and Pakistan—account for just over half of the world's irrigated land. The top 10 countries collectively account for two thirds of the world total. (See Table 3–1.) Many nations, including China, Egypt, India, Indonesia, and Pakistan, rely on irrigated land for more than half of their domestic food production. Because irrigated farms typically get higher yields and can grow two or three crops per year, the spread of irrigation has

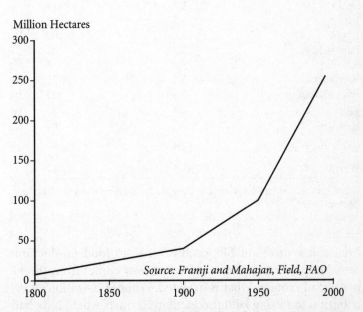

Million Hectares

Source: Framji and Mahajan, Field, FAO

Figure 3–1. Growth of World Irrigated Area, 1800–1995

Table 3–1. Irrigated Area in the Top 20 Countries and the World, 1995

Country	Irrigated Area	Share of Cropland That is Irrigated
	(million hectares)	(percent)
India	50.1	29
China	49.8	52
United States	21.4	11
Pakistan	17.2	80
Iran	7.3	39
Mexico	6.1	22
Russia	5.4	4
Thailand	5.0	24
Indonesia	4.6	15
Turkey	4.2	15
Uzbekistan	4.0	89
Spain	3.5	17
Iraq	3.5	61
Egypt	3.3	100
Bangladesh	3.2	37
Brazil	3.2	5
Romania	3.1	31
Afghanistan	2.8	35
Italy	2.7	25
Japan	2.7	62
Other	52.4	—
World	255.5	17

SOURCE: U.N. Food and Agriculture Organization, *1996 Production Yearbook* (Rome: 1997).

been a key driver in this century's rise in food production: some 40 percent of the world's food now comes from the 17 percent of cropland that is irrigated. Without irrigation's contribution to raising land productivity, farmers would have had to convert a much larger area of forest and grassland into cropland in order to produce today's harvest.[3]

The story of irrigation's advance over the last 150 years reveals not only how irrigation projects grew in number and scale to boost food production, but also how they were used to open up large new parts of the planet for human settlement and economic development. Irrigation became a tangible manifestation of a culture that viewed reclaiming deserts as human destiny, that lauded major feats of engineering and greater human control over nature, and that believed part of the business of government was building large water projects with public funds. The social and political milieu in which irrigation spread and flourished varied from one country to another, but these basic beliefs underpinned its advance in nations as different as China, Egypt, India, the former Soviet Union, and the United States. Recognizing irrigation's cultural basis not only sheds light on how irrigation progressed to its present state, it can help anticipate the forces of change that are shaping its future.

Irrigation Scales Up in India

The British are responsible for much of irrigation's advance during the nineteenth century. On the Indian subcontinent, then under British rule, they attempted waterworks of a whole new scale. Having honed their skills building railways and barge canals back home, British civil engineers began turning South Asia's rivers out of their banks and into massive new irrigation canals in an effort to safeguard the region from drought and famine. These activities, however, did not take place on a blank slate. British intervention disrupted many indigenous, ecologically sound, small-scale irrigation systems that had functioned successfully for centuries. (See Chapter 9.)

The British began building canals in India in 1817, and for the next two decades focused mainly on upgrading the water systems already in place. Then, after a severe famine in

1837–38, they initiated new and larger schemes aimed at greatly expanding the area under irrigation. Among the early projects was the Ganges Canal, which when it opened in 1857 was the largest in the world.[4]

Ironically, serious flaws in the Ganges Canal's initial design and construction spurred important advances in irrigation science. The principal architect, Sir Proby Thomas Cautley, had designed the canal's slope using standard hydraulic formulas developed by French scientists in the late eighteenth century. Those formulas aimed at ensuring that water moved fast enough to get distributed throughout the canal system, but not so fast as to scour and damage the banks and channels. For a scheme the size of the Ganges Canal, however, the formulas proved inappropriate, leading Cautley to design the canal's slope too steep. Within a short time, serious bank erosion and channel silting threatened to undermine the whole scheme. It took several years of trial and error to shore up the system. Out of these corrective efforts, however, came hydraulic formulas more suitable for the large-scale engineering schemes that would mark irrigation's new age. With slight modifications, these are still in use today.[5]

If modern irrigation has a birthplace, it is almost certainly the Punjab, the "Land of Five Rivers," situated in what is now northwest India and eastern Pakistan. Here, meltwater from the western Himalaya feed the Sutlej, Beas, Ravi, Chenab, and Jhelum Rivers, which all join the Indus River for its journey across a gently sloping sandy plain to the Arabian Sea. (See Figure 3–2.) These rivers annually carry some 175 cubic kilometers of water—twice the flow of the Nile—but the volume varies greatly by season. The rivers run high during the spring and summer, when they swell with snowmelt and rain, but drop off greatly during the winter months. The British viewed achieving better control of this flow as the ticket to turning the Punjab into a lucrative breadbasket.

British engineers began diverting irrigation water from the

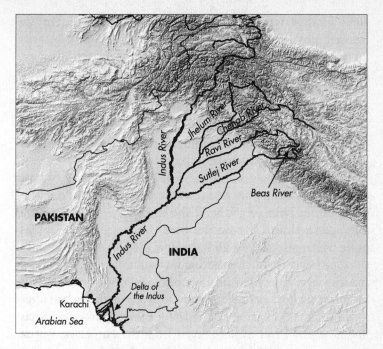

Figure 3–2. Rivers of the Punjab

Ravi and Sutlej Rivers during the first half of the nineteenth century, and then greatly stepped up the pace of canal-building during the century's latter half. Between 1870 and 1900, the area under irrigation in the Punjab tripled to 3 million hectares. As the twentieth century opened, the Punjab was already producing a surplus of food for export.[6]

Along with new design principles for large canal schemes, the British in India developed mechanized control technologies that transformed irrigation into a much larger enterprise. In particular, engineers equipped low-level dams called barrages with sliding gates that could raise the height of a river to the level needed for water to enter an irrigation canal. In this way, canals could be supplied with water even when the river level was low, which was often the time when irrigation was needed most. The irrigation water could then flow by gravity to farmers' fields. Gates at the head of the canal controlled how much irrigation water was admitted. In times of flood, the canal gates could be closed to protect land downstream.

These regulating structures allowed for much greater control of rivers than had been possible before. They enabled the British to bring the major tributaries of the Indus under control. Then early in the twentieth century, engineers turned to the larger challenge of the Indus itself. They built a barrage with 66 regulating gates and constructed seven major canals—three on the river's right bank and four on the left. When the barrage and canal headworks were completed in 1932, they were the largest of their kind in the world.[7]

One big piece of engineering was still to come in the Indus Basin—large reservoirs able to store some of the vast quantity of snowmelt for later use. As early as 1906, engineers had identified what they considered an ideal site for a dam—a gorge cut by the Sutlej River as it emerged from the Himalaya. But its construction would have to wait. Although engineers had built many other dams in India and elsewhere, they did not yet know how to build a dam big enough for this site. The answer eventually came from the United States, which pioneered the use of concrete in building dams of unprecedented height and mass. Indian engineers benefited from the experience gained by U.S. engineers during the construction of Hoover Dam on the Colorado River in the 1930s; construction of the Bakhra Dam on the Sutlej River began about mid-century.[8]

Throughout the Indian subcontinent, a combination of public and private initiatives spurred irrigation's advance. In addition to the large public schemes built by the British in the Punjab and elsewhere, Indian farmers themselves invested in groundwater wells and small storage reservoirs called tanks. By 1900, India had 13.4 million hectares under irrigation, an area larger by two thirds than the entire world's irrigated area 100 years earlier. Public schemes accounted for just over half of this development, and private investments for the remainder.[9]

In 1901, India's First Irrigation Commission called for a major investment program to quicken irrigation's spread. This initiative included the construction of public canal and tube-well projects, as well as assistance to private irrigators through low-interest loans, grants, and hydrologic surveys. By the mid-1940s, some 28 million hectares were under irrigation, more than double the area in 1900. With the subcontinent's partitioning after independence in 1947, about 30 percent of this irrigated land became Pakistan's and 70 percent became India's. A century of persistence, engineering hubris, and public and private enterprise had created an irrigation-based society dependent on a degree of water control the world had never seen before.[10]

Going West

As the British were embarking on their engineering schemes in the Punjab, Mormon settlers half a world away were building irrigation canals to transform their new home in the valley of the Great Salt Lake, in what would become the U.S. state of Utah. They were by no means the first irrigators in North America. The Hohokam, other native Indians, and Spanish settlers had been watering dry parts of the American West for centuries prior to the arrival of white Anglo-Saxon immigrants. (See Chapter 2.) Indeed, by 1800, Spanish communities had built more than 160 *acequias*, or irrigation canals, along the upper sections of the Rio Grande in what is now New Mexico.[11]

The Mormons, however, possessed a mission and zeal to master nature that took irrigation to a new and larger scale in the western hemisphere. Their hierarchical social organization and group discipline were reminiscent of the ancient irrigation societies. Church leaders controlled the design and location of new settlements, and allocated land and water to faithful followers. Their canals and rudimentary dams gradually became more sophisticated. When the transcontinental railroad opened in Utah in 1867, the Mormons began irrigating and marketing sugar beets and other cash crops, and plowed the profits back into more land development. By 1890, 43 years after the Mormons arrived in the valley below the Wasatch Mountains, more than 100,000 hectares of land were under irrigation in Utah. Although much smaller in scale than what the British had built in the Indus basin, the Mormon projects ushered in the age of large-scale, commercial irrigation in the United States.[12]

Early Mormon activities set the stage for other private irrigation ventures, and by about 1875 the center of irrigation innovation and development had shifted to Colorado and California. Eastern investors, including *New York Tribune* editor Horace Greeley, pumped money into land development near the confluence of the Cache la Poudre and South Platte Rivers in Colorado, experimenting with new forms of social organization and creating new settlements like Union Colony, precursor to the town of Greeley. In California, settlers focused mainly on mining after gold's discovery in 1848, but the wealth-creating potential of irrigated farming in this sundrenched place was not long overlooked by land speculators and agricultural entrepreneurs. By 1890, California had more than 400,000 hectares under irrigation—13 percent more than in Colorado and nearly four times more than in Utah.[13]

By this time, a new breed of irrigation enthusiasts was expounding the virtues of western land reclamation. They talked about irrigation in passionate terms—as a cause, a calling, and a destiny more than as a rational and economical way of expanding agricultural production. Among the biggest boosters was a former newspaper writer named William E. Smythe, who claimed that he "had taken the cross of a new crusade." Irrigation, he said, "not only makes it possible for a civilization to rise and flourish in the midst of desolate wastes; it shapes and colors that civilization after its own peculiar design. It is not merely the lifeblood of the field, but the source of institutions." Smythe's exuberance was countered by the cautionary views of John Wesley Powell, who had studied the West's irrigation possibilities as director of the Geological Survey. Although strongly in favor of western reclamation, Powell stated clearly that irrigation's potential in the West would be greatly constrained by a shortage of water, a view time would validate.[14]

As the nineteenth century came to a close, most of the land reclamation that could be accomplished with available technology and capital had been done. By this time, a system of water rights called prior appropriation had spread throughout the West, allocating water on a "first in time, first in right" basis. State offices existed to protect established water rights, and special irrigation districts had been formed to administer projects. As in the Punjab, however, there was a piece of engineering missing if irrigation was to fulfill the promise more and more people seemed to believe it offered for the American West: big dams to store water. River flows during much of the prime growing season were naturally low, and irrigators with senior water rights had already claimed most of them. Newcomers could only get water if a way was found to capture and store some of the heavy spring snowmelt that would otherwise run out to sea. Dams were the logical solution, but for-profit investment companies generally saw them as too risky and capital-intensive.[15]

Few believed that the federal government should directly take on the work of reclaiming the West through irrigation; instead, its role should be to offer assistance in the form of surveys, land grants, and the like. As late as 1890, Powell told a Senate committee that the actual work of reclamation should be left to private enterprise. The same year, another irrigation supporter, Senator William Stewart, told his congressional colleagues that although great irrigation systems had been built in

India and elsewhere "under monarchical or despotic rule" and with government funds, "there is no necessity for the United States to engage in such expenditures. If the opportunity is furnished to the people of this country, they will reclaim these desert lands so far as reclamation is necessary."[16]

Nevertheless, with irrigation expansion stalled in the hands of private and state interests, a consensus gradually emerged that the federal government should finance large dams and canals to further the reclamation and settlement of the West. More and more engineers had become aware of what the British had accomplished in India, and, with patriotic eagerness, they were ready to work similar wonders in the western deserts. In his inaugural address in 1901, President Theodore Roosevelt delivered a ringing endorsement of the idea, claiming that "great storage works are necessary to equalize the flow of streams and to save the flood waters. Their construction has been conclusively shown to be an undertaking too vast for private effort."[17]

Not long after, in June 1902, Roosevelt signed into law the National Reclamation Act, launching a new phase not only of irrigation expansion but of water development and use throughout the West. Some of the earliest schemes included dams of unprecedented size. The Salt River Project in Arizona, not far from where the Hohokam irrigation society had faded several centuries before, included Roosevelt Dam, the first of the nation's "high" dams. Completed within a decade of the Reclamation Act's passage, this dam ultimately turned more than 100,000 hectares of Arizona desert into productive cropland.[18]

As engineers set their sites higher, projects got bigger and bigger. In 1912, a California engineer, Joseph Lippincott, presaged the fate of the Colorado River when he noted: "We have in the Colorado an American Nile awaiting regulation, and it should be treated in as intelligent and vigorous a manner as the British government has treated its great Egyptian prototype."[19]

Whether the Colorado's treatment was intelligent is highly debatable, but that it was vigorous is indisputable. The construction of Hoover Dam in the 1930s broke all engineering records to date. Some 220 meters high, Hoover was the first of a generation of structures that became known as "super-dams"—those more than 150 meters high. In addition to the giant Bhakra Dam in the Indus basin, they also include Longyangxia Dam in China's Yellow River basin; Grand Coulee Dam on the Columbia in the northwestern United States, with a spillway more than twice the height of Niagara Falls; and Glen Canyon Dam on the Colorado, Hoover's upstream neighbor.

River after river came under Bureau of Reclamation control. In its first 50 years of existence, the Bureau (earlier called the Reclamation Service) constructed 173 dams, including the four highest and largest in the world. Federal reservoirs were able to store 63 billion cubic meters of water, equivalent to the average annual flow of three-and-a-half Colorado Rivers. By 1952, Bureau projects were irrigating 2.7 million hectares of cropland—just over a quarter of the nation's total irrigated area of 10 million hectares.[20]

Although impressive, the federal reclamation program actually brought less land under irrigation during the first half of the twentieth century than private and local initiatives had brought under irrigation during the last half of the nineteenth. The latter had brought the West's irrigated area to about 3.2 million hectares in 1902. Irrigation expansion had slowed by the century's turn for a good reason: not only were new, river-based irrigation projects becoming too large for private initiatives, they were becoming uneconomical in general. Reclamation would prove more difficult and costly than its boosters had envisioned.[21]

As Bruce Babbitt, Secretary of the Interior in the Clinton administration, describes it: "The Bureau became part of an extraordinarily powerful political force composed of the U.S. Congress, local interests, and a hungry bureaucracy. This coalition elected Westerners to Congress by promising to dam every single stream in the region, paid for with a continuous flow of tax dollars from people east of the Mississippi River. Thus did we create and subsidize a welfare state in the West, under the paternal guidance of the Bureau of Reclamation."[22]

The Transformation of the Nile

If British-ruled India gave birth to modern irrigation, and the United States imbued it with grandiose purpose and greatly scaled-up river engineering, the Egyptians gave irrigation perhaps its greatest challenge: sustaining tens of millions of people in a country that gets virtually no rain. All it takes is a flight into Cairo to see what the ancient Greek historian Herodotus meant when he called Egypt "the gift of the Nile." A sea of desert sand surrounds a narrow green strip of life on both sides of the river. Along with the Nile's fertile delta near the Mediterranean Sea, this band of watered earth has supported Egypt's civilization for millennia. For this arid nation, the Nile River is quite literally a lifeline.

Although Egypt ranks fourteenth in irrigated area today, its 5,000 years of continuous irrigated agriculture gives its modern experience with irrigation elevated importance. Perhaps more than any other country, Egypt has been driven by its population growth and extreme aridity to pursue a greatly intensified degree of water control. Its irrigated agriculture today differs markedly from the basin irrigation that sustained its people for millennia. (See Chapter 2.)

Prior to 1800, Egypt's cultivated area and population size had both probably peaked early in the first century A.D. During this time, Egypt supplied the Roman Empire with vast quantities of grain. Although estimates for ancient Egypt vary greatly, its maximum population early this millennium was probably 5 million. Early in the nineteenth century, however, the population started to climb at an unprecedented rate. Since flood-based basin irrigation could support crop production for only a third of the year, the number of Egyptians threatened to exceed the level that this traditional cultivation system could support. In response, Egypt began to turn to new methods of water engineering and control that fundamentally altered the ecological underpinnings of Nile valley agriculture.[23]

Napoleon was the first to suggest engineering schemes that could make more extensive use of the Nile's flood waters. Little happened, however, until closer to the mid-nineteenth century, when Muhammad Ali ordered the construction of a series of diversion dams across the Nile at the head of the fan-shaped delta, about 20 kilometers north of Cairo. Completed in 1861, this effort marked the onset of modern irrigation in the Nile valley. A series of other dams were built in subsequent decades, including the first Aswan Dam, completed by British engineers in 1902 and enlarged twice by 1934.[24]

British-built irrigation works transformed Egypt's irrigated agriculture from a seasonal system to a perennial one. What the *shaduf* and waterwheel had accomplished as a small-scale supplement to basin irrigation during ancient times, large dams and diversion canals did on a much grander scale. For two thirds of the twentieth century, Nile valley agriculture was a hybrid of traditional flood-based irrigation and modern perennial irrigation, although moving in the direction of the latter. As early as 1928, water scholar E.H. Carrier wrote in *The Thirsty Earth* that "disturbing factors have already begun to be manifest" in Egyptian agriculture. He wrote of the risks of waterlogging, which in Egypt's dry climate would lead to salinization. He cautioned that the silt that had replenished the fertility of the floodplain for millennia was now largely trapped behind the Aswan Dam. Trends in cotton yields already suggested a decline in soil fertility.[25]

By 1966, Egypt's population had climbed to about 30 million—a 10-fold increase in roughly a century and a half. The dams and reservoirs built by then were able to store about 9 billion cubic meters, just over 10 percent of the Nile's average annual flow. But with the introduction of thirsty cash crops like cotton and the rapid rise in population, 10 percent was not enough. With great fanfare and Soviet technical and financial assistance, President Gamal Abdel Nasser oversaw the construction of the High Dam at Aswan. "In antiquity, we built pyramids for the dead," he said. "Now we will build pyramids for the living."[26]

Completed in 1970 and located about 7 kilometers upstream from the old Aswan Dam, the High Dam towered at 111 meters and created a reservoir able to store nearly two full years of the Nile's average annual flow. The Aswan High Dam was a monument to Egypt's national pride as much as a solution to the desert nation's water challenges. To convey the degree to which even those who questioned the wisdom of the dam fell into line, one government official cited a verse from Omar Khayyám's Rubáiyát: "When the King says it is midnight at noon, the wise man says behold the moon."[27]

Many Egyptians who remember the terrible drought of the early 1970s claim that the High Dam paid for itself in a matter of years by helping avert a massive famine. A repeat performance in the mid-1980s emphasized the point. Egypt's irrigated area has climbed to nearly 3.3 million hectares, and officials hope to increase this area to 4.6 million hectares over the next two decades. But hard choices about how to best use the Nile's waters lie ahead. Egypt's population will soon reach 70 million, and is projected to rise to 115 million by 2050. Already, the nation ranks among the world's largest grain importers, and its political relations with upstream Ethiopia are strained over sharing the Nile's waters. (See Chapter 7.) It remains to be seen how sustainable a system underpins modern Egyptian society.[28]

Boom—and Bust?

Irrigation's story during the last half of the twentieth century divides into two parts—the boom years from 1950 to 1980, and the slowing of irrigation's expansion since then. During the first period, large government investments, major financial support from international donors and lenders, and the spread of new and better pumping technologies fostered irrigation's worldwide spread. Population grew rapidly during this time, demanding faster-paced agricultural growth to maintain adequate food supplies.

This period coincided both with the heyday of international dam-building and the Green Revolution—the package of high-yielding seeds and fertilizers that led to dramatic gains in agricultural output in many developing countries. Especially in the many parts of Asia with short wet seasons and long dry seasons, irrigation became an essential component of the Green Revolution package. Without a reliable water supply, farmers would simply not invest in expensive seeds and fertilizers.

In country after country, large new irrigation schemes came on-line during the third quarter of the twentieth century. The Chinese, who had already developed substantial irrigation in the well-watered south part of the country, turned their attention to the more problematic Yellow River basin and the north China plain. China suffers more than most countries from a severe imbalance between its population size and its water endowment: the nation has 21 percent of the world's people but only 7 percent of its renewable fresh water.[29]

Even worse, that water is distributed very unevenly both in time and space. Because of its monsoon climate, China gets most of the year's rainfall during the summer and early autumn. In Beijing, for example, 85 percent of the year's precipitation falls between June and November. In addition, four fifths of China's river runoff occurs in the southern part of the country, while just one fifth occurs in the north. Yet more than 60 percent of China's arable land is in the central and northern regions, and most of it requires irrigation to be highly productive.[30]

At the founding of the People's Republic in 1949, China's irrigation base was depressed from years of war, but still totaled an impressive 19.5 million hectares, about the same as India's at that time. Communist Party Chairman Mao Zedong put millions of peasants to work shoring up and expanding the nation's agricultural infrastructure. Over the next 35 years, the nation built more than 83,000 reservoirs, sunk 2.3 million wells, repaired or constructed 177,000 kilometers of dikes, and brought 29 million more hectares under irrigation. China's rural laborers carried out much of this work with little more than shovels, chisels, wheelbarrows, and locally made explosives.[31]

The new government also lost little time in constructing waterworks in three major river basins—the Huai, the Hai, and the Yellow. The latter proved particularly difficult from the start, prompting Chairman Mao to climb a hill outside Zhengzhou to issue a proclamation that "work on the Yellow River must be done well." The People's Victory Irrigation District in the river's lower reaches was completed in 1952, the first of more than 100 irrigation districts that would be built in Henan and Shandong provinces over the next several decades. By 1990, the Yellow River was irrigating nearly 4.4 million hectares, nearly a third as much as the Yangtze River to the south.[32]

As electricity became more widespread, and as pumping and well-drilling technologies improved, China turned to groundwater as a source of irrigation, particularly in the dry north China plain. The number of wells nationwide climbed steeply—from 110,000 in 1961 to nearly 2.4 million in the mid-1980s. Collectively, groundwater wells irrigate 8.8 million hectares, about 18 percent of the nation's total irrigated land. All told, China's irrigated area climbed 2.5-fold between 1949 and 1995—from 19.5 million hectares to nearly 50 million. (See Figure 3–3.)[33]

In India and Pakistan, as in China, the proliferation of groundwater wells accounted for much of irrigation's expansion during the last half-century. While government canal-building in India nearly doubled the area under surface irrigation between 1950 and 1985, the most impressive growth

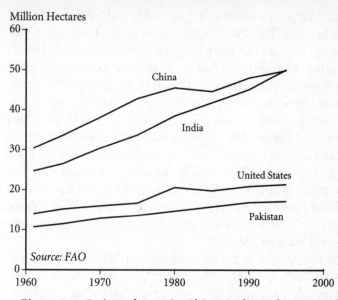

Figure 3–3. Irrigated Area in China, India, Pakistan, and the United States, 1961–95

was in groundwater development: the area irrigated by tube-wells ballooned from 100,000 hectares in 1961 to 11.3 million hectares in 1985, mostly as a result of private investment. In neighboring Pakistan, groundwater became the most rapidly growing source of irrigation water from the mid-1960s through the 1980s. A public program of tubewell development failed miserably, but private groundwater investments climbed steeply, with the total number of tubewells rising from about 25,000 in 1964 to nearly 360,000 in 1993.[34]

In the Indus River basin, now divided between India and Pakistan, irrigation engineers completed the work begun by the British more than a century earlier. With the construction of two massive storage dams—Tarbela on the Indus and Mangla on the Jhelum—and a corresponding extension of surface canals, the Indus Basin Irrigation System became the world's largest contiguous irrigation network, covering 14 million hectares. This basin alone now accounts for nearly 6 percent of the world's total irrigated area. All told, Pakistan's irrigated cropland reached 17.2 million hectares in 1995. Closely paralleling trends in China, India's irrigated area rose 2.4-fold between 1950 and 1995, to 50 million hectares.[35]

Both the former Soviet Union and the United States, which round out the top five irrigators during the last half-century, expanded their irrigated lands substantially between 1950 and 1980. The Soviet Union, though blessed with abundant water resources overall, was cursed with an unfortunate distribution of those resources: 84 percent of its river runoff flowed into the Arctic and Pacific Oceans, away from the major population centers and arable lands.[36]

A top priority of Soviet leaders was to expand irrigation in regions with favorable temperature conditions for crop production, but where rainfall was either insufficient or unreliable. They concentrated on two key areas—the Central Asian republics, which accounted for about 40 percent of Soviet irrigated area prior to the nation's breakup, and the southeastern European region, including parts of Russia and Ukraine. With irrigation water drawn from rivers feeding into the Aral Sea, Moscow turned the deserts of central Asia into a major cotton-growing region. Between 1940 and the early 1980s, Soviet cotton production quadrupled and the nation became the world's second largest cotton producer, accounting for about 20 percent of global production. Irrigation in the southeastern European areas helped safeguard this important grain-producing region from drought.[37]

Except in Central Asia, large-scale schemes played a comparatively small role in the Soviet Union's irrigation development. The majority of irrigated farms relied on local projects, such as pumping irrigation water from nearby rivers or tapping underlying groundwater supplies. Massive water development did take place, however, in central Asia, where irrigated area had expanded by the late 1980s to 7 million hectares—more than double Egypt's current total. Large diversions from the Amu Darya and Syr Darya, the twin rivers that feed the Aral Sea, supplied most of the irrigation water. (These diversions have cost the sea two thirds of its volume and caused unprecedented ecological destruction, as described in Chapter 5.) In all, Soviet irrigated area nearly doubled between 1970 and 1990, reaching almost 21 million hectares. With the country's breakup, Russia and Uzbekistan now rank among the world's top 20 irrigators, with 5.4 million and 4 million hectares, respectively.[38]

In the United States, irrigated land doubled from 10 million hectares in 1950 to 21 million hectares in 1995, as individual farmers sank untold numbers of groundwater wells, large state-funded projects came on-line, and one large federal project after another was completed. Under the leadership of the indomitable Floyd Dominy, Commissioner of the Bureau of Reclamation from 1959 to 1969, the share of western irrigated land supplied wholly or partially by Bureau water rose to a peak of nearly 25 percent. During his tenure, the Bureau also became a training ground for engineers from many developing countries. "Everywhere I went in India," Dominy boasts, "they thought I was the second coming of the Buddha."[39]

By the early 1990s, the Bureau of Reclamation had built more than 190 projects and was supplying about one third of all the surface water used for irrigation nationwide. As in many parts of the world, however, much of the irrigation boom came from private groundwater development. In a striking bit of good fortune, the U.S. Great Plains—which straddles the 100th meridian, the nation's transition zone from rain-fed to irrigated agriculture—is underlain by a vast reserve of groundwater in a geologic formation called the Ogallala. One of the planet's great aquifers, it spans portions of eight states, covers 453,000 square kilometers, and—prior to development—held 3,700 cubic kilometers of water, a volume equal to the annual flow of more than 200 Colorado Rivers. After World War II and the introduction of powerful centrifugal pumps, Great Plains farmers began tapping this water on a large scale, first in the aquifer's southern regions of northwest Texas and western Kansas, and then gradually further north into Nebraska. Today, the Ogallala waters one fifth of the nation's irrigated land.[40]

Irrigation projects are spread throughout the United States, but three regions—California, the Pacific Northwest, and the Great Plains—account for more than half of the nation's total irrigated area. Because of its fabulously productive rain-fed lands in the East and Midwest, the United States is much less dependent on its irrigated land than the world's other major irrigators. Of the top 20 countries in irrigated area, only Russia and Brazil have a smaller share of their cropland under irrigation than the United States does—4 and 5 percent, respectively, compared with 11 percent in the United States. Yet, as is true everywhere, this land is disproportionately valuable, yielding 38 percent of the total U.S. crop value.[41]

During the last two decades, irrigation's steady boom has begun to wane. Between 1970 and 1982, global irrigated area grew at an average rate of 2 percent a year. But between 1982 and 1994, this rate dropped to an annual average of 1.3 percent. Over the next 25 years, the global irrigation base is unlikely to grow faster than 0.6 percent a year, and even this may turn out to be optimistic. World population growth has also slowed in recent years, but not as much as irrigation has. Per capita irrigated area peaked in 1978 and has fallen 5 percent since then. By 2020, per capita irrigated area will likely be 17–28 percent below the 1978 peak. (See Figure 3–4.)[42]

Irrigation has simply begun to reach diminishing returns. In most areas, the best and easiest sites are already developed. Bringing irrigation water to new sites is more difficult and costly. In India and Indonesia, for example, the costs (in inflation-adjusted terms) of new irrigation schemes have more than doubled since about 1970; in Thailand, they have risen 40

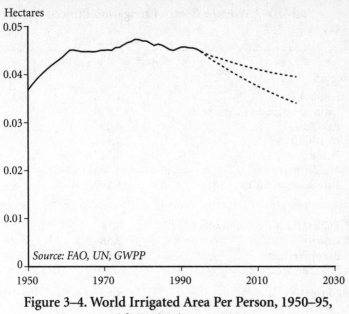

Figure 3–4. World Irrigated Area Per Person, 1950–95, with Projections to 2020

percent, and in the Philippines, more than 50 percent.[43]

A 1995 study by the World Bank of more than 190 Bank-funded projects found that irrigation costs now average just over $4,800 per hectare. Average costs vary greatly from one region to another, from a low of $1,370 in South Asia to a high of $18,269 in sub-Saharan Africa. (See Table 3–2.) When the Bank adjusted these costs to account for projects that were incomplete because of construction delays and other implementation problems, costs rose markedly higher in some areas. In sub-Saharan Africa, they soared to more than $31,000 per hectare.[44]

In many ways, Africa has missed out on the modern irrigation age: only 6 percent of the continent's cropland is under irrigation, compared with 37 percent of Asia's. (See Table 3–3.) Moreover, 70 percent of Africa's irrigated land is in just five

Table 3–2. Average Costs of Irrigation Projects[1]

	Average Cost (dollars per hectare)
Region	
South Asia	1,370
Latin America and Caribbean	3,923
East Asia	4,291
North Africa	4,911
Europe	4,743
Middle East	5,062
Sub-Saharan Africa	18,269
Type of Project	
Pump (mostly groundwater)	3,766
Gravity	5,584
Rehabilitation	1,633
New construction	7,740
All Projects	4,837

[1]Data are for 191 projects funded by the World Bank.
SOURCE: William I. Jones, *The World Bank and Irrigation* (Washington, DC: World Bank, 1995).

countries—Egypt, Madagascar, Morocco, South Africa, and Sudan. Very little of Africa's irrigation benefits the sub-Saharan region, where deep poverty and food insecurity persist. (See Chapter 9.)[45]

Combined with historically low crop prices, the high costs of irrigation have made investments in new irrigation schemes increasingly uneconomical. Worldwide irrigation lending by the four major donors—the World Bank, the Asian Development Bank, the U.S. Agency for International Development, and the Japanese Overseas Economic Cooperation Fund— peaked in the late 1970s and dropped by nearly half over the next decade. Governments saddled with debt burdens and budget deficits became increasingly unwilling or unable to finance new projects. China, India, Indonesia, and several

Table 3–3. World Irrigated Area by Region, 1995

Region	Irrigated Area	Share of World Total[1]	Share of Cropland That is Irrigated
	(million hectares)	(percent)	(percent)
Asia	175.4	69	37
North and Central America	30.1	12	11
Europe	25.1	10	8
Africa	12.3	5	6
South America	9.8	4	8
Oceania	2.6	1	5
World	255.3	101	17

[1]Column does not add to 100 due to rounding.
SOURCE: U.N. Food and Agriculture Organization, *1996 Production Yearbook* (Rome: 1997).

other Asian governments cut back their irrigation investments substantially during the 1980s.[46]

Moreover, many of the world's irrigation systems have been poorly maintained. After many years or decades of use, 50–70 percent of them are badly in need of repair. On average, rehabilitating existing systems costs one fifth as much as constructing new ones, and makes good economic sense in most cases. But with limited funds available for water projects overall, the need to repair and upgrade aging networks will constrain irrigation expansion.[47]

One country has publicly stated that its era of government-built irrigation projects is over. In 1993 the Commissioner of the U.S. Bureau of Reclamation, Daniel P. Beard, announced a dramatic departure from his agency's 91-year mission of reclaiming deserts and controlling rivers. In his *Blueprint for Reform*, Beard stated plainly that "Federally-funded irrigation water supply projects will not be initiated in the future. This decision reflects the marginal national economic benefit from new irrigation investment and the need to focus limited Federal funding on increasing efficiency and remediating adverse impacts of existing projects." More colloquially, Beard said that his agency was "stuck working on yesterday's issues. We were reading off an old script."[48]

Most governments are not yet at the point of calling a halt to their irrigation work. A number still have substantial plans for publicly funded irrigation expansion, including Brazil, China, Egypt, India, and Turkey. But the trend toward declining public investments is unmistakable. The new look of irrigation will increasingly be private rather than public, and smaller rather than bigger.

As the twentieth century ends, the 150-year modern irrigation age is winding down. The cultural, economic, and political forces that shaped this era and underpinned the impressive rise of irrigation-based societies around the world are being realigned. On the economic front, irrigation expansion in many areas has reached the point of diminishing returns. At the same time, new political constituencies and cultural values are reframing the debate about large water projects. From western urbanites and recreation enthusiasts in the United States to environmentalists and human rights advocates in India, new voices have begun to question irrigation's ends and means. As the next chapters will show, preserving the world's irrigation assets will be no small task. Managing and expanding those assets to meet the food and livelihood challenges ahead will take fundamentally new approaches.

4

RUNNING OUT

On February 7, 1997, as the Chinese were celebrating their traditional Spring Festival, something strange happened to their "mother river," the Yellow, the cradle of China's civilization. It stopped flowing. From Jinan, the capital of Shandong Province, to the river's mouth, the channel went bone dry. It stayed that way for 130 days straight. Then the river started flowing again, but it stopped at least eight more times that year. All told, the river was dry in its lower reaches for nearly two thirds of the year, a record 226 days.[1]

This was not the first year the Yellow River had dried up, but in 1997 the stoppage happened earliest and lasted the longest. The previous two years had each set records as well, with the river failing to reach the sea for 133 days in 1996 and 122 days in 1995. Although a mid-1990s drought worsened the situation, the principal cause of the river's condition was heavy human claims on its flow. The Yellow has run dry every year this decade, with the dry section often stretching 600 kilometers, from Henan Province all the way to the river's mouth.

After a monitoring station in Shandong Province first recorded zero flow in 1972, the river stopped flowing in 19 of the next 25 years.[2]

Half a world away, in Deaf Smith County, Texas, a similarly ominous bell tolled in 1970. A groundwater well drilled in 1936 to irrigate farmland in the Texas High Plains went dry. More than three decades of pumping had caused the underground water table to drop 24 meters. As in China, this dry well in Texas signaled much worse to come. Water tables were dropping across a wide area, which, when energy prices shot up a few years later, would force farmers to close down thousands of wells.[3]

Many of the world's most important food-producing regions are simply running out of water for irrigation. Across large areas, farmers are pumping groundwater faster than nature is replenishing it, causing a steady drop in water tables. Just as a bank account dwindles if withdrawals routinely exceed deposits, so will an underground water reserve dry up if pumping chronically exceeds recharge. During the last two decades, groundwater depletion has spread from isolated pockets to large areas of irrigated cropland. The problem is now widespread in central and northern China, northwest and southern India, parts of Pakistan, much of the western United States, North Africa, the Middle East, and the Arabian Peninsula. Groundwater overpumping may now be the single biggest threat to irrigated agriculture, exceeding even the buildup of salts in the soil.

Many of the planet's major rivers are suffering from overexploitation as well. Just like falling water tables, overtapped rivers signal that there is little additional water to supply to farmers. In Asia, where most of the world's population growth and additional food needs will be centered, many rivers are completely tapped out during the drier part of the year, when irrigation is so critical to food production.

China's Conundrum

Among the world's overtapped rivers, none is more telling or troubling than China's Yellow River, known to the Chinese as the Huang He. The Yellow is China's second largest river, carrying 58 billion cubic meters (bcm) of water each year. Typically, 60 percent of this flow occurs during the rainy season, July through October, when little irrigation is needed. Half of the yearly irrigation demand occurs between March and June, when the river carries a fourth of its annual flow. In addition, the Yellow carries so much sediment that at least a third of its flow is used to transport the heavy silt load toward the coast. So just 37 bcm are available to meet the basin's water demands, which have risen rapidly: since 1950, diversions from the Yellow by the 10 provinces that share it have increased fivefold. The river now supplies 140 million people and 7.4 million hectares of irrigated farmland.[4]

Total water demand in the basin already exceeds the usable supply by 10 percent, and it is projected to climb 45 percent by 2030. Because of limited water, these higher demands will not be satisfied, raising the possibility of severe crop losses as downstream provinces—especially Shandong, the last in line—lose out on irrigation water. The farmers of Shandong Province produce roughly one fifth of China's wheat and one seventh of its corn. A quarter of the province's cropland, nearly 2 million hectares, gets irrigation water from the Yellow River.[5]

In 1995, the drying up of the Yellow River slashed crop output by 2.7 million tons. In 1997, crop losses were placed at 13.5 billion yuan ($1.7 billion), which at $200 per ton implies a reduction of some 8.5 million tons. At a September 1997 meeting of experts to discuss the Yellow River's condition, State Council vice premier Jiang Chunyun called attention to the "contradictions between economic and social progress on the one hand and the environment and resources" on the other. "If we fail to take practical and effective countermeasures, [the] water supply crisis along the Huang He valley will gradually worsen with unthinkable consequences."[6]

Among the linchpins in China's strategy for alleviating shortages in the Yellow River basin are two of the largest and most technically complex water schemes ever undertaken—the massive Xiaolangdi Dam and diversions from the Yangtze River basin in the south to the north China plain.

The Xiaolangdi project is the latest and largest in a long line of attempts to tame the Yellow River by storing more of its floodwaters. The Yellow is by far the siltiest of the world's major rivers. It emerges from the highly erosive Loess Plateau carrying 1.6 billion tons of sediment, more than three times the load of the Yangtze, which carries 16 times more water. Historically, the Yellow carries 1.4 billion tons of its sediment all the way to the sea, dropping the remaining 200 million tons in its lower reaches as it loses speed. As a result, the river bed rises about 10 centimeters a year, and now sits 4–7 meters higher than the surrounding cities and plains. Over the last 50 years, the Chinese government has allocated vast amounts of money and labor to reinforce dikes along 1,300 kilometers of riverbanks and to construct flood control dams upstream.[7]

With assistance from the Soviet Union, the Chinese completed construction of a large dam at Sanmenxia (Three Gate Gorge) on the Yellow River in 1960. The impounded reservoir filled with silt so quickly that the ancient capital of Xian downstream was soon threatened with flooding. Between 1962 and 1973, engineers twice redesigned the dam to provide for better flushing of sediment. Nevertheless, sediment already claims about 40 percent of Sanmenxia's reservoir capacity, greatly limiting its hydropower and flood control potential. Unfortunately, Sanmenxia is no exception. The reservoir behind Yangouxia Dam on the upper Yellow lost nearly a third of its storage capacity before it came into full operation.[8]

Despite this track record, the Chinese hold out great hope

for Xiaolangdi, not only to safeguard the lower basin from floods, but to store more precious Yellow River water for use by cities and farms. Situated in Henan Province and slated for completion in 2001, Xiaolangdi will be China's second biggest dam, after the more well known Three Gorges on the Yangtze River. It will crest at 154 meters, and its reservoir is designed to store 12.6 billion cubic meters of water—a fifth of the river's total annual flow. At a projected cost of about $3 billion, including the resettlement of 181,000 people, Xiaolangdi is the largest foreign-financed project undertaken in China, with World Bank loans and credits totaling $570 million.[9]

The Chinese hope that by storing massive quantities of water and trapping millions of tons of silt, Xiaolangdi will solve the siltation, flooding, and water shortage problems in the Yellow River basin in one gigantic stroke. Like other efforts to tame the Yellow, however, the benefits may prove short-lived, if not elusive: According to some estimates, the reservoir behind Xiaolangdi will fill with silt in about 30 years, and long before that its usefulness for flood control and water supply will be greatly diminished.[10]

China's other ambitious engineering scheme is a diversion from the Yangtze River in the south to increase water supplies in the north. Communist Party Chairman Mao Zedong reportedly dreamed up the idea on a trip down the Yangtze in 1958. For the last four decades, the diversion scheme has been a topic of study, debate, and controversy, with media reports variously indicating that it has received a go-ahead, that it is still under review, or that it is only a possibility.

Engineers have sketched out three possible routes, and work is apparently going ahead on two of them. (See Figure 4–1). The eastern one would extend the old Grand Canal, which now stops in southern Shandong province, in order to bring Yangtze water to Tianjin, an important commercial city on the north China plain. The middle route, also under construction, would siphon water from a tributary of the Yangtze

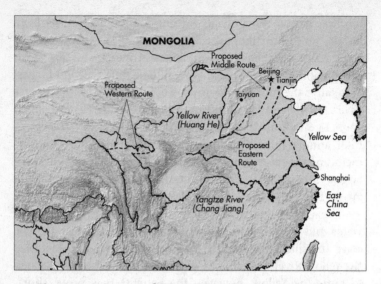

Figure 4–1. Proposed River Diversion Routes in China

and deliver it to Beijing. The western route, the least likely of the three to be completed, would transfer flows from upper Yangtze tributaries directly to the upper reaches of the Yellow River, where the two rivers are the closest.[11]

If all three diversions were completed, they would transfer 45–60 bcm of water a year, roughly in the range of the Yellow River's total annual flow. They would require boring tunnels through mountains, building aqueducts stretching over 1,000 kilometers, and constructing power plants to pump the water uphill. The projected costs, which typically turn out to be underestimates on such large projects, total $30 billion.[12]

Not surprisingly, the project's size, cost, and complexity have made it enormously controversial. According to some accounts, there is more internal opposition to the diversion scheme than to Three Gorges Dam. Nonetheless, the Chinese leadership is determined to proceed. The drying up of the Yellow River in recent years has renewed a sense of urgency to

solve the north's water problems. With water demands projected to far outstrip supplies in Beijing and Tianjin, however, little if any of the diversion water would be available to expand
irrigation.[13]

Draining Rivers Dry

China is by no means the only country with overtapped rivers.
The Ganges and Indus in South Asia, the Nile in northeast
Africa, the Amu Darya and Syr Darya in Central Asia, the Chao
Phraya in Thailand, and the Colorado in southwestern North
America are each now so dammed and diverted that for parts
of the year little or none of their fresh water reaches the sea.
Along with mounting ecological damage, these diminished
flows portend worsening water shortages that may make it dif-
ficult to sustain irrigated agriculture, much less expand it.[14]

Prior to construction of the Aswan High Dam in Egypt, for
example, some 32 billion cubic meters of Nile water reached
the Mediterranean Sea each year, equal to 38 percent of the
river's total annual flow. After the dam was built and more
river water was diverted for crop production, flow to the sea
dropped dramatically—to some 6 bcm a year. As irrigated
area continued to expand, outflow to the sea fell even further,
to about 3 bcm a year in the mid-1980s. Today, the annual
volume of fresh water reaching the Mediterranean totals just
1.8 bcm, and virtually all of it is released to the sea during
the few winter months when crops do not require as much
irrigation water.[15]

Despite nearing the limits of its Nile supplies, Egypt continues to plan for expanded irrigation. In 1997, President
Hosni Mubarak gave the go-ahead for a controversial scheme
called the Toshka or New Valley project, which would divert
Nile water upstream from Lake Nasser and transfer it hundreds
of kilometers to irrigate some 200,000 hectares of Egypt's
southwestern desert. Mubarak also opened a new canal to carry

Nile water beneath the Suez Canal, aimed at eventually irrigating 250,000 hectares of the Sinai desert. Stepped-up reuse of agricultural drainage water will allow for some additional irrigation expansion as well. But Egypt has insufficient supplies to expand water-intensive agriculture much further, and even some of its current activities are ruffling feathers in upstream countries, especially Ethiopia. (See Chapter 7.)[16]

Egypt also faces serious long-term threats posed by the sinking of the Nile delta. Most deltas subside from the weight of their own sediment, but under natural conditions this subsidence is counteracted by new silt deposits brought in by the river. The Nile transports an average of 110 million tons of silt each year, much of it fertile soil washed down from the Ethiopian highlands. For thousands of years, the river carried 90 percent of this silt to the coast, where it replenished the delta. It deposited the remaining 10 percent on the Nile floodplain, creating fertile soil for Egyptian farmers. After the first small dams were built by the British about a century ago, some of this sediment remained trapped in reservoirs, and the delta stopped growing. Following completion of the High Dam at Aswan, and the trapping of virtually all of the silt in Lake Nasser, the delta has actually been retreating. Borg-el-Borellos, a former delta village, is now 2 kilometers out to sea.[17]

Two of the great rivers of South Asia—the Ganges and the Indus—now have virtually no outflow to the sea during the dry season. Along with more than 1 million hectares of irrigated cropland, the Indus delta region contains the world's fifth largest mangrove forest. The mangroves harbor a wide variety of wildlife, including several endangered species of birds and dolphins. Today, because of water development upstream, no fresh water reaches the delta during the dry season from October to March. The delta environment, and mangroves in particular, has deteriorated markedly from rising salinity, the decline of nutrients, erosion, and land subsidence.[18]

Similar problems have arisen in the Ganges basin, which is

shared by Nepal, India, and Bangladesh. India's heavy diversions upstream during the dry season leave almost nothing in the river for Bangladesh, much less to reach the river's natural outlet in the Bay of Bengal. Besides constraining irrigation in Bangladesh, the lack of fresh water flowing into the Ganges delta has caused the rapid advance of a saline front across its western portion, which is damaging valuable mangroves and fish habitat—important resources for the local economy. The Sundarbans, one of the world's largest mangrove forests, is also home to the threatened Bengal tiger.[19]

In Thailand's Chao Phraya basin, water demands already exceed available supplies. Flows for navigation are consistently less than optimum. Shortages of river water in Bangkok have kept the city dependent on overpumping groundwater, a practice that is causing portions of the city to sink. A team of World Bank specialists warned in 1993 that "as nonagricultural demands increase, and if no additional water becomes available, supplies for dry season cropping will tend to decline."[20]

Shrinking Groundwater Reserves

Overtapped rivers are easy to see and the consequences fairly apparent. Groundwater overpumping, though hidden from view, may be an even more serious problem.

In India, the situation is so severe that in September 1996 the nation's highest judicial body, the Supreme Court, directed one of India's premier research centers to examine the problem of declining groundwater levels. The National Environmental Engineering Research Institute (NEERI), based in Nagpur, found that "overexploitation of ground water resources is widespread across the country" and that water tables in critical agricultural areas are sinking "at an alarming rate."[21]

The situation is particularly serious in the states of Punjab and Haryana, India's principal breadbaskets. Village surveys found that water tables are dropping 0.6–0.7 meters per year in parts of Haryana and half a meter per year across large areas of Punjab. In the western state of Gujarat, 87 out of 96 observation wells showed declining groundwater levels during the 1980s, and aquifers in the Mehsana district are now reportedly depleted. Overpumping in this western coastal state has also caused salt water to invade freshwater aquifers, contaminating drinking water supplies. In the southern state of Tamil Nadu, underground water levels have dropped up to 30 meters since the 1970s, with aquifers in the Coimbatore district now depleted.[22]

Because of the rapid proliferation of groundwater wells, now numbering at least 6 million, and the failure to regulate pumping adequately, several key agricultural states in India face a severe and worsening water imbalance. The NEERI study shows that water demands exceed available supplies in nine states. Their combined water deficit totals just over 100 billion cubic meters a year, a volume equal to the average annual flow of 1.2 Nile Rivers. (See Table 4–1.) Since it takes about 1,000 cubic meters of water to grow one ton of grain, this deficit suggests that a substantial portion of India's total grain production is being produced unsustainably. David Seckler, Director General of the International Water Management Institute in Sri Lanka, estimates that a quarter of India's grain harvest could be in jeopardy from groundwater depletion.[23]

Water budgets are particularly out of balance in Gujarat, Haryana, Karnataka, Maharashtra, Punjab, and the desert state of Rajastan. Farmers cannot overpump groundwater indefinitely. At some point, when their costs get too high or their well yields drop too low, they will be forced to choose among several options. They can take irrigated land out of production, reduce the number of crops they plant on each plot of land, switch to less water-intensive crops, or adopt more-efficient irrigation practices. Apart from shifting out of thirsty nonfood crops like sugarcane or cotton, improving efficiency is the only option that can sustain food production while lowering water use. Yet investments in

Table 4–1. Water-Deficit States in India, Mid-1990s

State	Annual Water Deficit
	(billion cubic meters per year)
Rajasthan	32.6
Maharashtra	22.0
Gujarat	16.0
Haryana	14.2
Karnataka	12.7
Punjab	4.0
Other	2.8
Total	104.3

SOURCE: National Environmental Engineering Research Institute, "Water Resources Management in India: Present Status and Solution Paradigm," Nagpur, India, undated (circa 1997).

efficiency are minuscule relative to the challenge at hand.

Besides constraining future food production, groundwater overpumping is widening the income gap between rich and poor in some areas. As water tables drop, farmers must drill deeper wells and buy larger pumps to lift the water to the surface. The poor cannot afford these technologies. In parts of Punjab and Haryana, for example, wealthier farmers have installed submersible tubewells at a cost of about 125,000 rupees ($2,950). As the shallower wells dry up, some of the small-scale, poorer farmers end up renting their land to the larger well owners and becoming laborers on these larger farms.[24]

In neighboring Bangladesh, groundwater use is about half the level of natural groundwater replenishment on an annual basis. During the dry season, however, when most irrigation occurs, heavy pumping causes many wells to go dry. On about a third of Bangladesh's irrigated area, water tables routinely drop below the suction level of shallow tubewells during the dry season. Although the aquifers recharge and water tables rise again during the monsoon season, farmers still run out of water when they need it. Again, poor farmers—who cannot afford to deepen their wells or buy bigger pumps—suffer the most.[25]

In China, which is roughly tied with India for the most irrigated land, groundwater conditions are as unsettling as the diminished river flows. North China is in a chronic water deficit, with groundwater overpumping amounting to some 30 billion cubic meters a year. Of the three major river basins in northern China, the Hai is always in deficit, the Yellow is almost always in deficit, and the Huai is occasionally in deficit. This northern and middle China plain produces roughly 40 percent of China's grain. Across a wide area, the water table has been dropping 1–1.5 meters a year, even as water demands continue to increase.[26]

Modeling work by Dennis Engi of Sandia National Laboratories in New Mexico suggests that the water deficit in the Hai basin could grow by more than half between 1995 and 2025, even assuming that the Chinese complete at least part of the Yangtze River diversion. The Sandia model projects that the deficit in the Yellow River basin could balloon by 190 percent. The combined deficit in these two river basins could more than double over the next three decades, from 27 bcm in 1995 to 55 bcm in 2025.[27]

As in India, overpumping groundwater cannot fill these gaps indefinitely. Like their Indian counterparts, China's farmers will either take land out of irrigated agriculture, switch to less thirsty crops, or irrigate more efficiently. How they respond will make a big difference to China's grain outlook: its projected 2025 deficit for the Hai and Yellow River basins is roughly equal to the volume of water needed to grow 55 million tons of grain—14 percent of the nation's current annual grain consumption and more than a fourth of current global grain imports.[28]

In Pakistan and the United States, portions of irrigated agriculture are operating in water deficits as well. In Pakistan's province of Punjab, its leading agricultural region, groundwater pumping exceeds recharge by an estimated 27 percent. The Ogallala aquifer in the United States, which waters one fifth of U.S. irrigated land, is being depleted at a rate of about 12 bcm a year. Net depletion so far totals some 325 bcm, a volume equal to the annual flow of 18 Colorado Rivers. More than two thirds of this depletion has occurred in the Texas High Plains.[29]

Driven by falling water tables, increased pumping costs, and historically low crop prices, many farmers who depend on the Ogallala have already abandoned irrigated agriculture. At its peak in 1978, the total area irrigated by the Ogallala in Colorado, Kansas, Nebraska, New Mexico, Oklahoma, and Texas reached 5.2 million hectares. Less than a decade later, this area had fallen by nearly 20 percent, to 4.2 million hectares. Projections made for a long-range study of the region back in the mid-1980s suggested that more than 40 percent of its peak irrigated area would come out of irrigation by 2020; if this happens, another 1.2 million hectares will either revert to dryland farming or be abandoned over the next two decades.[30]

California is overdrafting groundwater at a rate of 1.6 billion cubic meters a year, equal to 15 percent of the state's annual net groundwater use. Two thirds of this depletion occurs in the Central Valley, which supplies about half of the nation's fruits and vegetables. Annual pumping varies considerably between wet years and dry years, but the long-term trend is one of depletion. Added to the Central Valley's salinity and toxic drainage problems, this degree of groundwater overpumping portends a contraction of farming in this prime agricultural region.[31]

In North Africa and the Arabian Peninsula, much groundwater use depends on nonrenewable aquifers—those that get little or no replenishment from rainfall today. Pumping water out of such aquifers depletes the supply, much as extracting oil depletes an oil reserve.

No country has a more dramatic recent history of groundwater depletion than Saudi Arabia. After the oil embargo of the

1970s, the Saudis realized that they were vulnerable to a retaliatory grain embargo. The government responded by launching a major initiative to make the nation self-sufficient in grain. It heavily subsidized land, equipment, and irrigation water, and by buying crops from farmers at several times the world market price, it encouraged large-scale wheat production in the desert. From a few thousand tons in the mid-1970s, the annual grain harvest grew to a peak of 5 million tons in 1994. Saudi water demand at this time totaled nearly 20 billion cubic meters a year, and 85 percent of it was met by mining nonrenewable groundwater. Saudi Arabia not only became self-sufficient in wheat; for a time, it was among the world's wheat exporters.[32]

But this self-sufficiency would not last. Crop production came crashing down when King Fahd's government was forced to rein in expenditures. Within two years, Saudi grain output fell by 60 percent, to 1.9 million tons in 1996. Today Saudi Arabia is harvesting slightly more grain than in 1984, the year it first became self-sufficient, but because its population has grown so much since then, now numbering more than 20 million, the nation has again joined the ranks of the grain importers.[33]

Moreover, the Saudis' massive two-decade experiment with desert agriculture has left the nation much poorer in water. In its peak years of grain production, the nation ran a water deficit of 17 bcm a year, consuming more than 3,000 tons of water for each ton of grain produced in the hot, windy desert. At that rate, underground water reserves would have run out by 2040, or possibly sooner. In recent years, the annual depletion rate has probably dropped back to the level of the mid-1980s, but the Saudis are likely still running a water deficit of about 6 bcm a year.[34]

Africa's northern tier of countries—from Egypt across to Morocco—also relies heavily on fossil aquifers, with estimated depletion running at 10 bcm a year. Nearly 40 percent of this depletion occurs in Libya, which is now pursuing a massive water scheme rivaled in size and complexity only by China's Yangtze River diversion. Known as the Great Man-Made River Project, the $25-billion scheme pumps water from desert aquifers in the south and transfers it 1,500 kilometers north through some 4,000 kilometers of concrete pipe. The project is so big that the London-based Pipeline Industries Guild honored it with the 1996 award for the most "significant contribution to land-based pipeline technology."[35]

The brainchild of Libyan leader Muammar Qaddafi, the artificial river was christened with great pomp and ceremony in late August 1991. As of early 1998, it was delivering 146 million cubic meters a year to Tripoli and Benghazi. If all stages are completed, the scheme will eventually transfer up to 2.2 billion cubic meters a year, with 80 percent of it slated for agriculture. As in Saudi Arabia, however, the greening of the desert will be short-lived: some water engineers say the wells would likely dry up in 40–60 years.[36]

While some observers call the scheme "madness" and a "national fantasy," foreign engineers involved in the project have questioned Qaddafi's real motives. In a December 1997 *New York Times* article, engineers point out that the pipelines are 4 meters in diameter, big enough to drive a truck or move military troops through. Every 85 kilometers or so, engineers are building huge underground storage areas that apparently are more elaborate than needed for holding water. The master pipeline runs through a mountain where Qaddafi is reported to be building a biological and chemical weapons plant. Subsequently, others have scoffed at the possibility of any military motive, noting, for example, that the pipeline system has no air vents.[37]

The upshot of this survey of groundwater use is that many important food-producing regions are sustained by the hydrologic equivalent of deficit financing. Irrigators are drawing down water reserves to support today's production, racking up large water deficits that at some point will have to be balanced. Just how much of the world's irrigated land is watered by over-pumping groundwater is difficult to estimate. Annual water depletion in India, China, the United States, North Africa, and the Arabian Peninsula adds up to about 160 billion cubic meters a year. (See Table 4–2.)

If inclusion of the rest of the world raised this volume by 25 percent—a not unreasonable assumption—water deficits worldwide would amount to 200 billion cubic meters a year. The vast majority of this overpumped groundwater is used to irrigate grain, which suggests that on the order of 180 million tons of grain—about 10 percent of the global harvest—is being produced by depleting water supplies. These findings raise an unsettling question: If so much of irrigated agriculture is oper-ating under water deficits now, where are farmers going to find the additional irrigation water needed to satisfy the food demands of the more than 2 billion people projected to join humanity's ranks by 2030?

Table 4–2. Water Deficits in Key Countries and Regions, Mid-1990s

Country/Region	Estimated Annual Water Deficit
	(billion cubic meters per year)
India	104.0
China	30.0
United States	13.6
North Africa	10.0
Saudi Arabia	6.0
Other	unknown
Minimum Global Total	163.6

SOURCE: Various references cited in the text and author's estimates.

Dams—Salvation or Sorrow?

Globally, there is still water we could tap on a sustainable basis. Much of it, however, is not accessible. About a fifth of all the water running to the sea each year is too remote to supply any cities or farming regions. About half runs off in floods. Much of the remainder occurs in regions where abundant rainfall makes irrigation unnecessary.[38]

For most of this century, water managers focused on building dams to capture some of the floodwater, store it in a reservoir, and deliver it to farmers and city dwellers. Some 40,000 large dams (those over 15 meters high) now block the world's rivers—up from 5,000 in 1950. Small dams number in the 800,000 range. Collectively, reservoirs worldwide are capable of storing on the order of 6,600 cubic kilometers of water—about a fifth of the annual volume of floodwater heading to the sea each year. Considerably less water gets delivered to farms and cities, however, since dams and reservoirs are also managed to generate electricity, reduce flooding, and enhance river navigation.[39]

During the heyday of dam-building, from the 1950s to the mid-1970s, about 1,000 large dams came on-line every year. Engineers naturally selected the most favorable and least costly sites first. As these were developed, new projects became more complex and expensive. By the 1980s, rising costs were joined by heightened concern about the social and environmental damage done by large dams, and the pace of construction slowed. By the early 1990s, engineers were completing about 260 large dams a year—still a large number, but just a quarter as many as during the boom years.[40]

There is good reason to think the pace of dam-building will slow even further. Over the last 15 years, a strengthening worldwide anti-dam movement has forced the cancellation or indefinite postponement of a number of large dams in a variety of countries. One of the most protracted and highly publicized dam battles has been over Sardar Sarovar, the centerpiece of a large scheme in western India's Narmada River valley that includes 30 large dams, 135 medium ones, and 3,000 small ones. The intensity of the battle over Sardar Sarovar reflects how deeply polarized the positions of those supporting large dams and those opposing them have become. Out of this confrontation and the many other dam controversies of the last decade, however, may come a constructive reevaluation of where, when, and under what conditions additional large dams are justified.[41]

Like most large dam projects, Sardar Sarovar promised a multitude of benefits—hydropower for electrification, drinking water for some 40 million people, and irrigation water for 1.8 million hectares of cropland. With India's population growing by 18 million a year, the need for additional food, energy, and drinking water is undeniable. Whether Sardar Sarovar is the proper way to meet these needs, however, became a major point of disagreement. Critics pointed to distorted estimates of the project's benefits and incomplete renderings of its costs. Moreover, because the Indian state of Gujarat, where the dam is located, would reap most of the expected benefits while 90 percent of the lands to be submerged by the reservoir are located in Madhya Pradesh and Maharashtra, the project posed serious questions of fairness.[42]

For more than a decade, a creative strategist and social activist named Medha Patkar has led a persistent campaign against Sardar Sarovar. Supported by nongovernmental groups around the world, this effort has focused on the lack of proper environmental assessments to ensure the project's sustainability and the inadequacy of resettlement and compensation plans for the tens of thousands of "oustees," mostly poor tribal villagers forced to leave their homes to make way for the reservoir and canals. In 1990, Japan became the first major foreign donor to withdraw its support for Sardar Sarovar.[43]

Further protests by Patkar and the Narmada Bachao Andolan (NBA), or Save the Narmada Movement, led the

World Bank, which had approved $450 million for the project, to initiate an independent review of the environmental impacts and resettlement plans. The resulting report was a scathing critique that underscored numerous flaws and inadequacies, and recommended that the Bank "take a step back" from the project. In March 1993, the Bank announced that it was withdrawing its support. The Indian government vowed to proceed on its own.[44]

That same year, monsoon floodwaters blocked by the dam wall, then 44 meters high, inundated village lands, washing away the homes and possessions of 40 families who had refused to move. The NBA responded with more hunger strikes, nonviolent protests, and threats of *jal samarpan* (self-sacrifice by drowning). The organization also filed a case with the Supreme Court in New Delhi.[45]

In 1994, the government of Madhya Pradesh announced that it had neither the money nor the land needed to resettle the project oustees properly, and that it wanted the planned height of the dam to be lowered. With more fasting and protesting, NBA encouraged this upstream state to call for a complete halt to the dam pending progress on resettlement, which state officials did in late 1994. In January 1995, NBA won a key victory when the central government in Delhi forced the state of Gujarat to halt construction of the dam wall, which at that time was just under half the planned final height. In February 1999, India's Supreme Court issued an interim order allowing the dam wall to be raised an additional 5 meters, but it has yet to make a final determination on the project's fate.[46]

Supporters of Sardar Sarovar see tragedy in this outcome, pointing to the enormous cost of delaying the project's hoped-for benefits. While acknowledging the potential for waterlogging, salinization, and other environmental damages, they see these as resolvable management issues. They point out that the 40 million people expected to benefit from the project dwarf the 240,000 people who will be harmed.[47]

Few of the project's supporters, however, dismiss the serious issues of resettlement and compensation for those who lose their homes and livelihoods. A 1992 commentary by the U.S.–based Winrock International Institute for Agricultural Development pointed out, for example, that compensation is both necessary and achievable—at least in theory. It suggested that since the amount of farmland that would be lost because of the dam is only 5 percent of the irrigated area that would be created, and since the number of people who would be harmed is less than 1 percent of the number expected to benefit, even a small tax on the beneficiaries would generate enough revenue to compensate the losers.[48]

In reality, however, compensation for the Sardar Sarovar "oustees"—which has focused on resettlement—has fallen short. Moreover, since governments as well as the World Bank have notoriously poor track records when it comes to compensating the victims of large dam projects, critics have learned not to trust that plans on paper will be adequately implemented. Harald Frederiksen, a water engineer formerly with the World Bank who supports construction of Sardar Sarovar, notes that a key lesson from this experience is the need for officials to "act promptly to remedy any flaws in the resettlement and environmental measures" associated with a dam project, and that in the case of Sardar Sarovar, "the states did not respond fully or in a manner matching the urgency."[49]

Historians will no doubt see the battle over Sardar Sarovar, as well as others like it over the last decade, as a turning point. Meeting the legitimate food, water, and material needs of the rising global population will almost certainly require that additional dams be built to store more floodwater. The challenge, however, is for governments, donors, private companies, and others proposing to build additional dams to analyze properly the full range of benefits and costs—including economic, environmental, health, and social costs—and to ensure effective compensation to those harmed by these projects. A thorough evaluation of alternatives—including conservation, efficiency improvements, and small-scale options—is also critical to the acceptance of any dam as the preferred approach to meeting human needs.

Until these failings are corrected, organized opposition will lead to the delay and cancellation of more and more projects. With the rapid spread of Internet access and electronic mail, anti-dam activists are now able to orchestrate well-coordinated international campaigns against what they consider the most environmentally or socially destructive dam projects. As Patrick McCully of the International Rivers Network in Berkeley, California, notes in his book *Silenced Rivers*, "Public protests are provoked by just about every large dam that is now proposed in a democratic country. The international dam industry appears to be entering a recession from which it may never escape."[50]

All major sides of the big-dam debate took an important step forward in April 1997, when they agreed to the establishment of an independent commission to evaluate the "development effectiveness" of large dams and to assess "if and how they can contribute to sustainable development." Chaired by South Africa's Minister of Water Affairs and Forestry and a former human rights lawyer, Kader Asmal, the 12-member World Commission on Dams includes representatives of a diverse array of interests, perspectives, and expertise. If the work of the Commission helps set a course for the proper evaluation of dam proposals and the proper implementation of dam projects, it will have done a great service. The Commission is to announce its findings and recommendations in June 2000.[51]

Whatever the outcome of this process, the pace of dam-building will slow during the years ahead. World agriculture cannot expect to receive large increases in water supplies from additional dams, reservoirs, and river diversions. Moreover, even as it is getting more difficult to expand water supplies, various human activities are reducing the supplies already developed. For example, cutting trees in watersheds and paving over aquifer recharge areas often reduce the infiltration of rainwater into the soil. More water then runs off in floods rather than seeping into the subsurface and recharging groundwater.

In addition, even as new dams add to water storage capacity, siltation is continuously reducing it—as noted earlier, the Chinese have faced this problem at Sanmenxia on the Yellow River. Many of India's reservoirs are filling with silt much faster than expected. Nizamsagar reservoir in Andhra Pradesh lost more than 60 percent of its capacity over 40 years, one of India's worst cases. Pakistan's massive Tarbela Dam on the Indus River lost 12 percent of its live storage capacity within the first 18 years of operation.[52]

Worldwide, reservoirs are estimated to be losing storage capacity to sedimentation at a rate of 1 percent a year. If current global reservoir capacity is about 6,600 cubic kilometers, this rate suggests that some 66 billion cubic meters of storage are lost annually. Replacing this lost storage by building new reservoirs could easily cost $10–13 billion a year, assuming enough new reservoir sites could be found. If suitable sites did not exist, and engineers had to dredge the sediment out of existing reservoirs, costs could climb to $130–200 billion a year. Like salinization and groundwater depletion, the silting up of reservoirs is a quiet, creeping threat that is building to massive proportions.[53]

The Climate Wild Card

Last but not least, the buildup of carbon dioxide and other heat-trapping gases in the atmosphere confronts irrigated agriculture with the prospect of a changing climate. Like the joker in a game of cards, climate change has the potential to greatly alter the "game" of agriculture, but scientists do not know exactly how, when, or under what conditions this wild card will be played.

They do know that Earth's temperature will rise, which in turn will intensify the global hydrological cycle. A warmer atmosphere will hold more moisture, increasing global rates of evaporation and precipitation by some 7–15 percent. Rainfall patterns will shift, with some areas getting more precipitation and some getting less. River flows will change. Hurricanes and monsoons are likely to intensify, and sea level will rise from thermal expansion of the oceans and the melting of mountain glaciers and polar ice caps.[54]

Although the major climate models agree fairly well on global-scale changes, they are not finely tuned enough to predict what will happen regionally and locally. This uncertainty makes it difficult to plan wisely for new dams, reservoirs, and irrigation systems that are supposed to last a half-century or more. Most disturbing, in cases where climate change results in less rainfall, areas already at or near water limits may move into a long period of shortages.[55]

Many of the world's important irrigated areas depend on water from mountain snowmelt. These include much of the Indus and Ganges River basins in South Asia, the Aral Sea basin in Central Asia, the Colorado basin in the U.S. Southwest, and the Sacramento–San Joaquin valleys in California. Mountain snowpack acts like a reservoir, storing water in the winter and then releasing it during the spring and summer as the snows melt. The dams, reservoirs, and irrigation systems built in these regions are designed and operated with this pattern of river runoff in mind.[56]

These systems may be particularly at risk in a warmer world. With more precipitation falling as rain and less as snow, the volume of water stored naturally in mountain snowpacks will drop, and more winter precipitation will immediately become river flow. Moreover, the snowpack will melt earlier and faster, causing more risk of flooding in the spring and reducing the amount of water available just when irrigated agriculture needs it most—during the hot, dry summer.

John Schaake, a hydrologist with the U.S. National Weather Service, analyzed potential changes in the pattern of flow of the Animas River where it runs through Durango, Colorado. He found that a temperature increase of 2 degrees Celsius, with no change in precipitation, would have little effect on the total volume of annual runoff. The seasonal pattern of that runoff, however, would change greatly because of the reduced winter snowpack and its faster melting. Schaake's model showed that, compared with current runoff patterns, average runoff in January through March would increase by 85 percent, while in the critical months of July through September it would fall by 40 percent. Without more reservoirs to store the increased winter-spring runoff, serious shortages would likely occur in summer—just the time when competition for water is keen for irrigation, for hydroelectricity, and to keep rivers flowing enough to sustain fisheries, dilute pollution, and support recreational activities.[57]

If, for the sake of illustration, similar kinds of altered river flows around the world necessitate a 20-percent increase in reservoir storage capacity, some $200–400 billion in new investment could be required just to sustain the current irrigated area. Additional investment would be needed to expand irrigation systems to farming regions where rainfall becomes insufficient or too unreliable. In a warmer climate, increased evaporation in the spring would dry out the soil in some areas, leaving less moisture for evaporation and local rainfall in the summer. If even 2 percent of existing rain-fed land worldwide requires irrigation to remain productive, the climate price tag could rise by another $120 billion—assuming today's average cost of irrigation projects. Future costs would be considerably higher.[58]

Even if governments, international donors, and private investors could come up with such large sums, food security will likely be at risk for decades as planners, engineers, and farmers try to discern if the rainfall and runoff shifts that are occurring are long-term or temporary, and thus whether such investments are justified. Peter Gleick of the California-based Pacific Institute notes that "one of the most difficult climatic changes for most regions to handle would be increased variability—and thus more frequent extremes: higher peak floods, more persistent and severe droughts, greater uncertainty about the timing of the rainy reason. Few such changes would be beneficial."[59]

There will, of course, be countervailing positive influences. Higher carbon dioxide levels in the air have a fertilizing effect on many crops, boosting rates of photosynthesis. They also cause plants to narrow the opening of their stomata, which reduces water consumption. Because these physiological changes can lead to higher crop yields and lower crop water use, they partly explain why some scientists find little cause for concern about climate change's impacts on agriculture. But for crops to benefit from the fertilizing effects of carbon dioxide, they must have sufficient soil moisture; otherwise, potential yield gains will quickly turn to losses. For example, a lack of soil moisture during the flowering, pollination, and grain-filling stages of growth is especially damaging to maize, wheat, soybean, and sorghum. If soils dry out in late spring and summer, and irrigation water is not available or sufficient to make up the deficits in soil moisture, harvests will suffer.[60]

Some models that paint a fairly rosy picture about climate change's impacts on world agriculture fail to take into account this critical issue of water availability. In one study reported in the science journal *Nature*, for example, modelers assumed that "water supply for irrigation would be fully available at all locations under climate change conditions." In effect, the researchers assumed away a potentially big part of the problem.[61]

Although no one can say precisely how irrigated agriculture will be affected by global warming, a few things are fairly certain. First, the future will not be a simple extrapolation of the past. Most water planning for the future, however, is necessarily based on past trends. Engineers are designing dams for the twenty-first century based on twentieth-century hydrology data, even though river flows may change considerably as the climate warms. Second, for some period of time, reservoir and irrigation systems are likely to be poorly matched to altered rainfall and runoff patterns. This may leave farmers short of water during the dry summer months. And third, if and when the needed adjustments are made, they will be costly.

5

A FAUSTIAN BARGAIN

In the classic German legend, Faust makes a pact with the devil, surrendering his soul in exchange for 24 years of occult power on Earth. Near the end of the term, the devil comes to claim his soul. Faust could have saved himself by acknowledging and repenting his excesses, but he does not, and the devil drags him into the underworld.[1]

Our modern society may have inadvertently struck a Faustian bargain as well, in our case with nature. In return for transforming deserts into fertile fields and redirecting rivers to suit human needs, nature is exacting a price in myriad forms. Among the most threatening is the scourge of salt—the creeping, insidious menace that undermined the stability of several ancient irrigation societies, and that now places ours in jeopardy as well. All it takes is a look at southern Iraq, site of the earliest Mesopotamian civilizations, to be reminded that where salt claims a victory, it can be a very long-lasting one.

Mineral salts naturally occur dissolved in rain, rivers, and groundwater, and are bound up in soil particles as well. They include sodium, calcium, magnesium and potassium chlorides, sulfates, and carbonates. When farmers irrigate their crops, salts in the irrigation water get deposited in the soil. Even good-quality water typically has salt concentrations of 200–500 parts per million (ppm), the higher figure being the U.S. government's recommended limit for drinking water. If a farmer annually applies 10,000 tons of irrigation water to a hectare of crops, which is fairly typical, between 2 and 5 tons of salt will be added to that land every year. Unless these salts are flushed out, enormous quantities can build up over the course of years or decades.

Salts can enter the root zone from below, as well. As irrigation water seeps through the soil from farm fields and unlined irrigation canals, the underground water table rises. Over time, if this water is not drained away, the root zone becomes waterlogged, starving plants of oxygen. In drier climates, when the groundwater gets to within a meter or two of the surface, plant roots pull it up through the upper layers of soil. The water then evaporates, leaving the salts behind. These two sources of soil salinity—the irrigation water applied at the surface, and rising groundwater from below—often work in tandem. Gradually, the salt buildup reduces the land's productivity, causing crop yields to decline. Eventually, if the situation is not corrected, the land has to be abandoned, just as it was in ancient Mesopotamia.

Worldwide, one in five hectares of irrigated land suffers from a buildup of salts in the soil. Vast areas are losing productivity in China, India, Pakistan, Central Asia, and the United States. (See Table 5–1.) Soil salinization costs the world's farmers an estimated $11 billion a year in reduced income, and this figure is growing. Spreading at a rate of up to 2 million hectares a year, soil salinity is offsetting a good portion of the increased productivity achieved by expanding irrigation. Salt may well present as great a risk to modern society as it did to the ancients.[2]

Table 5–1. Salinization of Soils on Irrigated Lands, Selected Countries and the World, Late 1980s

Country	Irrigated Land Damaged by Salt	Share of Total Irrigated Land Damaged by Salt[1]
	(million hectares)	(percent)
India	7.0	17
China	6.7	15
Pakistan	4.2	26
United States	4.2	23
Uzbekistan	2.4	60
Iran	1.7	30
Turkmenistan	1.0	80
Egypt	0.9	33
Subtotal	28.1	21
World Estimate[2]	47.7	21

[1]Based on 1987 irrigated area figures to maintain consistency with the approximate time of the salinization estimates, except irrigated areas of Uzbekistan and Turkmenistan, which are taken from *FAO, 1996 Production Yearbook*. [2]Assumes same share of world irrigated area is affected as collectively is affected in the eight countries shown.

SOURCE: Adapted from F. Ghassemi, A.J. Jakeman, and H.A. Nix, *Salinisation of Land and Water Resources* (Sydney: University of New South Wales Press, 1995).

Salt in the Aral Sea Basin

The destruction of the Aral Sea in Central Asia ranks near the top of the world's environmental tragedies. This region may also harbor the world's biggest salt problem. It is a place of many bizarre sights. On a visit there in 1995, I saw a graveyard of ships rotting in the dried-up seabed. I stood on a seaside bluff outside the old port town of Muynak, but I saw no water—the coastline and the sea were 40 kilometers away. But to my eyes, the strangest and eeriest sight of all was the salt: vast areas of the land glisten white, like new-fallen snow.[3]

Several decades ago, Moscow central planners made a calculated decision to sacrifice the world's fourth largest lake in order to turn the deserts of Central Asia into a lucrative cotton-growing region. As noted in Chapter 3, increasing quantities of water were diverted from the region's two major rivers, the Amu Darya and Syr Darya, to supply an expanding area of irrigated land, which now totals nearly 8 million hectares. Since 1960, the sea has lost two thirds of its volume, the local fishing economy has virtually collapsed, and high rates of cancers, respiratory ailments, and many other diseases plague the local population.[4]

As tragic and costly as these consequences are, they may in the end be dwarfed by the mounting problem of salt. Losing a good portion of the region's irrigated agriculture, which employs millions of people, would cause widespread social and economic turmoil. But unless the salt problem is solved, this is precisely what could happen.

The movement of salt through a region's land and water is not always easy to measure and track. In the Aral Sea basin, as much as 120 million tons of salt may be in flux—more than in any other river basin in the world, including the Indus, with nearly twice as much irrigated land. As the Amu Darya and Syr Darya head toward the sea, numerous diversions deplete their flows, and salty farm drainage returns to their channels from the surrounding fields, causing salt concentrations to rise steadily. Salinity levels in the lower reaches of the Syr Darya, for instance, are six times higher than in its headwaters.[5]

As a result, downstream irrigators are adding enormous quantities of salt to their cropland. Moreover, they apply additional water to their fields during the noncrop season in order to push the accumulated salts down and out of the root zone before they plant their next crop. This extra "leaching" water not only increases their total water demand, it adds still more salt to the land.

To make matters worse, seepage from the heavy irrigation applications is recharging the underlying groundwater at a rate hundreds of times faster than natural levels. Among other things, this rapid recharge has mobilized very deep, highly saline groundwater, allowing it to mix into the groundwater column that now extends nearly to the soil surface. Evaporation of this extra-salty groundwater leaves large amounts of salt behind in the soil. This problem is particularly severe in the middle reaches of the two rivers, where drainage water flowing out of several large irrigation schemes is 5–10 times saltier than the incoming irrigation water.[6]

Managing the vast quantities of salt moving through the region's environment is tricky, to say the least, and even with a substantial portion of the basin's irrigated land equipped with drainage, the battle is not being won. In 1994, 28 percent of the basin's irrigated land had salt buildup severe enough to be lowering crop yields by 20–50 percent, compared with 23 percent four years earlier. In Karakalpakstan, the region of Uzbekistan most severely affected by the Aral Sea tragedy, 95 percent of the land has high salt levels. Not surprisingly, crop yields there have plummeted: between the mid-1970s and early 1990s, cotton and rice yields dropped by about a third.[7]

Even a rich country would have trouble orchestrating a workable solution to a salt problem this serious. In these young, economically struggling nations, the task is even more daunting. Turkmenistan actually expanded its irrigated area by 31 percent between 1990 and 1994 in an effort to boost its food self-sufficiency. Engineers did not equip the new irrigated lands with drainage, so the salt problem there may worsen. Even if governments install additional drainage facilities in the basin, a substantial area of irrigated land may need to be retired from crop production and dedicated to storing salt—possibly as much as 2 million hectares in Turkmenistan and Uzbekistan alone, a quarter of the basin's total irrigated area.[8]

Losing Ground in Pakistan

Pakistan has tackled its salt problems more vigorously than any major irrigator, but is still fighting an uphill battle. Geologically, the vast Indus plain was once covered by a shallow sea, into which surrounding rivers deposited their sediments. When the sea receded, salts were left behind in both the soil and the underlying groundwater. The natural weathering of rocks added substantial amounts of salt to the environment as well.[9]

When large-scale irrigation got under way during the mid-nineteenth century, the Indus basin environment already harbored a great deal of salt. In addition, the plain has a very gentle slope, which makes it difficult to drain. Together, the large amounts of salt and naturally poor drainage make irrigated agriculture a risky enterprise in this region, as the ancient Harappan civilization probably discovered 5,000 years ago. (See Chapter 2.)

At the time of independence in 1947, some 40,000 hectares of irrigated land in Pakistan were going out of production annually because of waterlogging and salt buildup. Very few irrigation canals had been lined, and with only about 3,700 kilometers of drains in operation, the underlying water table was rising steadily. Aerial photographs taken between 1952 and 1958 show that nearly 5 million hectares of land were highly salty, and nearly as large an area exhibited patches of salt. Field investigations confirmed the severity of the threat. By 1961, the government had drawn up a nationwide plan for combatting both waterlogging and salinity.[10]

Early on, the government's Salinity Control and Agricultural Reclamation Programme (SCARP) focused on installing vertical wells to pump groundwater in an effort to lower the water table below the danger zone. This pumped groundwater also provided a supplemental source of irrigation water. With assistance from the World Bank and other donors, the Pakistani government installed nearly 13,000 tubewells.[11]

Two big problems, however, doomed the effort. First, installation of the SCARP tubewells was fully subsidized by the government, but farmers were expected to pay back capital and operating costs. Since they had not been included in the program's planning, the farmers viewed the tubewells as the government's responsibility, and so many refused to pay for them. In addition, few made the water management improvements on their farms that were integral to the program's success.[12]

Second, use of the pumped groundwater to supplement irrigation supplies backfired. The water was of such poor quality that in some areas this practice worsened the very salt problem that the project was trying to correct. To top it all off, SCARP was very costly. At the program's height, funding for salinity control and reclamation accounted for 43 percent of the government's total water-related expenditures. By the mid-1980s, Pakistan was spending 60 percent more to operate and maintain public tubewells than it was to run the country's entire canal irrigation network. By 1990, the nation's cumulative costs for salinity control and reclamation had risen to nearly $1 billion.[13]

These high costs, combined with the tubewells' declining performance, led the government to devolve responsibility for groundwater development to the private sector. The idea was to phase out the SCARP tubewells in the areas with good groundwater, and let the private sector take over. By this point, private groundwater development had already taken off in Pakistan. As of 1993, farmers had installed more than 350,000 tubewells on their own, many of them in the agriculturally productive province of Punjab.[14]

This heavy groundwater pumping lowered the water table enough to slow the salt problem considerably; but the problem has by no means been solved. Moreover, groundwater use has reached unsustainable levels in many parts of the country. In the Indus basin as a whole, groundwater pumping is now estimated to exceed recharge by half.[15]

Studies by Jacob Kijne at the Sri Lanka–based International Water Management Institute show that excessive water use and continued salt buildup threaten the sustainability of significant portions of Pakistan's food base. Kijne calculated water and salt balances for three of the nation's irrigated areas—a canal system in the North-West Frontier Province, and two sites in the Punjab. At the first location, he found classic problems of waterlogging and salt buildup resulting from canal seepage, overirrigation, rising water tables, and lack of adequate drainage. At the Punjab sites, groundwater overpumping has lowered the water table, yet salt remains a serious threat because of the extensive recycling and reuse of salty groundwater pumped out of the upper layers of the aquifer—the same problem found during SCARP. Unless farmers adjust their agricultural practices, Kijne says, "further degradation of land and water resources is inevitable."[16]

Kijne's recommended solutions are a bitter pill to swallow for a country that gets more than 80 percent of its food from irrigated land and that is growing by 1 million people every three months. To combat the problems of the Punjab, the choices are to reduce the area planted in crops throughout the year, to shift the cropping pattern so that less area is planted in thirsty crops like rice, or to pursue some combination of these two options. For one of the Punjab sites, for example, Kijne calculates that making the system sustainable would require reducing the annual cropping intensity—a measure of the total area harvested in a year taking into account lands that grow more than one crop—from 130 percent to 93 percent, or, alternatively, planting substantially less rice and sugarcane.[17]

A Spreading Plague

Pakistan and the Aral Sea basin countries have two of the world's most intractable salt problems. But to varying degrees salt plagues all major irrigators, and many of the minor ones.

Salinity has challenged China's irrigated agriculture for many centuries. A book written in 1617, during the Ming Dynasty, included a prescription for salinity control, recommending that the water tables be kept deep in order to prevent salt buildup near the soil surface. Today, salinization affects some 6.7 million hectares, or 15 percent, of China's irrigated land—and is especially serious in parts of the Yellow River basin and the north China plain. Harold Dregne, a soil scientist at Texas Tech University, and Zhixun Xiong and Siyu Xiong of the Ningxia Academy of Agro-Forestry Sciences conclude that, given the extent and severity of the salinity problem, "the long-term potential for sustainable irrigation development in China's drylands seems to be gloomy."[18]

In India, 7 million hectares are estimated to be losing productivity because of salt buildup, and the problem is worsening. About 36 percent of the affected land lies in the plains of the Indus and Ganges Rivers, including portions of the agriculturally important states of Haryana, Punjab, Madhya Pradesh, and Uttar Pradesh. In Haryana, for example, which is an important breadbasket for India, the 15 billion cubic meters of water delivered by canal each year brings in more than 2 million tons of salt. Over an area of about 400,000 hectares, the water table has risen to within 3 meters of the surface—dangerously high. If corrective actions are not taken, this high-water-table zone threatens to spread to 2 million hectares of farmland.[19]

In the western United States, scientists and agricultural engineers continue to battle both salt and toxic elements in agricultural drainage water. Salinization affects 23 percent of U.S. irrigated land overall, and the proportion is substantially higher in several key western irrigated farming regions—including two thirds of the irrigated land in the lower Colorado basin and 35 percent of that in California. Salt buildup in the lower Colorado caused a foreign policy dispute during President John F. Kennedy's administration, when, in late 1961, salt levels in the Colorado River crossing into Mexico climbed to 2,700 ppm—more than five times drinking water quality, and a level that is damaging to many crops. The river's salinity naturally rises steadily as it flows toward the Gulf of California, but the key culprit in the incident was highly saline drainage discharging into the river from a federal irrigation district in southern Arizona. Outraged Mexican officials eventually got the United States to agree that river water crossing the border would be no more than 115 ppm saltier than that reaching Imperial Dam on the U.S. side.[20]

Subsequently, the U.S. Bureau of Reclamation constructed a large desalting plant at Yuma, Arizona, a $235-million insurance policy that the nation would meet its obligation to Mexico. (The plant has never fully operated because other, less expensive ways were found to meet the water quality commitments.) In addition, Congress passed legislation in 1974 authorizing a basinwide salinity control program. By 2015, some 19 projects are expected to remove more than 1.2 million tons of salt from the basin annually. Ranging in costs from $15 to $133 per ton of salt removed, most of the projects aim to reduce the volume of highly salty drainage water by lining canals, leveling cropland, and taking other measures that improve irrigation efficiency. Only time will tell whether the effort succeeds at substantially reducing the estimated $750 million in annual U.S. economic losses—which include not only lower crop productivity, but the cost of corrosion, extra water treatment, and other household and industrial damages—caused by salt in the lower Colorado River basin.[21]

California's Central Valley, the nation's premier fruit and vegetable basket, is another major U.S. region at risk. Much of the valley is underlain by poorly drained soils derived from marine sediments that are naturally high in salts. Heavy irrigation over the last several decades has dissolved these salts and moved them into shallow groundwater aquifers, where they concentrate because of naturally poor drainage. Some 1 million hectares in this region are affected by drainage and salt problems, which are particularly serious on the west side of the San Joaquin valley.[22]

Federal and state officials initially planned to manage the problem by building a 250-kilometer concrete canal to ship the salty farm drainage to the Sacramento–San Joaquin delta, just upstream from San Francisco Bay. The canal's first phase terminated near a place called Kesterson, which was supposed to serve as a regulating reservoir as the drainage moved north to the delta. Due to funding problems and concerns about possible pesticide contamination of the delta, however, the canal was never completed. As a result, Kesterson became a terminal reservoir for Central Valley drainage.

Kesterson's location on the Pacific Flyway made it prime waterfowl habitat, especially because the majority of the region's wetlands had already disappeared. Unaware of any possible harm, the U.S. Fish and Wildlife Service agreed to let the Bureau of Reclamation divert more of the fresh water flowing into Kesterson for irrigation purposes in exchange for the inflow of farm drainage from the Central Valley. In 1983, five years after the drainage began to arrive at Kesterson, it became clear what a tragic mistake this was. Biologists began to discover deformed, dying, and dead birds in droves.

It turned out that irrigation in the valley had washed not only salts out of the soil, but also selenium, a naturally occurring element that is essential in trace amounts but poisonous at high concentrations. Public outrage ensued when the *San Jose Mercury-News* ran a series of alarming color photographs of some of Kesterson's worst victims—including a tiny hatchling with no eyes, no feet, and only half a beak. Dubbed the "Three Mile Island of irrigated agriculture," Kesterson became a powerful symbol of irrigation's Faustian bargain.[23]

Following the Kesterson discovery, scientists found similar patterns of death and deformity at a number of other U.S. locations—including the Tulare Basin, also in California's Central Valley, and several other national wildlife refuges in the West. Fish and wildlife with dangerously high selenium levels have turned up in Colorado, Montana, New Mexico, Nevada, Oregon, Utah, and Wyoming, in addition to California. Yet there has been no concerted, coordinated effort to deal with selenium-laced agricultural drainage from western irrigation. A 1985 federal order to stop the flow of drainage to Kesterson led to the creation of numerous toxic evaporation ponds throughout the San Joaquin valley. Yet in the Tulare basin in the southern part of the valley, selenium-laced evaporation ponds are just about the only breeding habitat left for migratory birds, and rates of bird deformities there exceed those found earlier at Kesterson.[24]

A joint federal-state management plan issued in 1990 identified the need to construct clean waterfowl habitats to mitigate the toxic evaporation ponds, as well as to retire about 30,000 hectares of toxic and degraded valley farmland over the next 50 years. But mitigation merely buys time; it does not fix the fundamental unsustainability of the valley's agricultural practices. Failure to solve the region's salt and selenium problems ultimately threatens at least 200,000 hectares of irrigated land—not to mention birds and wildlife, and possibly human health as well.[25]

Getting Serious About Salt

Can we win the salt and drainage battle, or—like the ancient Sumerians—are we doomed to suffer irrigation's decline?

In theory, most salt problems can be solved. In reality, however, most remain unsolved. Engineers have devised many drainage techniques to control waterlogging and the salt buildup that comes with it. They can build horizontal drains on or below the soil surface, for example, or install vertical tubewells to pump out groundwater that rises too close to the surface.

Because waterlogging and salinity problems usually emerge many years after intensive irrigation has begun, however, installing drainage at the time irrigation systems are built is usually uneconomical. Often, drainage systems do not get built even when they are needed. Besides their expense, they lack the political appeal of a shiny new dam, and so they are prone to getting shifted down the list of priorities. In a gentle nod to what it calls the "drainage conundrum," a World Bank study of the performance of Bank-funded irrigation projects concluded that the Bank should not insist that its client countries construct drainage systems before they are needed, but that it should at least fund pilot schemes to identify the best solutions before large-scale schemes get under way.[26]

Besides the problem of getting drainage systems built, there is the issue of what to do with the drainage that is collected. Releasing it to a nearby river is a common practice. But this can severely degrade the quality of river water, potentially harming downstream ecosystems and water users. Dumping the drainage into a bay or other nearby body of water is another possibility, as was originally proposed for the San Joaquin valley drainage that ended up at Kesterson. Egypt releases farm drainage to the Mediterranean Sea. Pakistan is constructing a large drain to transfer irrigation drainage from the lower Indus basin to the Arabian Sea.[27]

Southern California's Imperial and Coachella Valleys ship much of their contaminated drainage to rivers flowing into the Salton Sea, a large inland water body created in 1905 and 1906 when a flooding Colorado River broke through its banks and flowed into an ancient dry lake bed about 240 kilometers southeast of Los Angeles. Because its only sources of inflow are highly polluted rivers and drainage water, the Salton Sea has become steadily more contaminated and is now about 25 percent saltier than the Pacific Ocean.[28]

Visitors to the area, especially in summer, are greeted by a stench that, when the wind is right, can waft all the way to Palm

Springs. Because of its location, the sea is a critical stopover on the Pacific Flyway. Some 380 bird species have been sighted there—more than anywhere else in North America except the Texas coast in spring. Birds and fish are dying by the thousands at the Salton Sea each year from epidemics and ailments of uncertain origin, but almost certainly a consequence of the salty, toxic soup the sea has become.[29]

Yet another option for disposing of drainage is to release it into ponds and let it evaporate. Especially where large quantities are involved, however, this method can require a considerable sacrifice of land. The managers of Australia's Murray River basin turned to this approach, in part because sending the drainage all the way to the sea was too expensive. In some areas, 3 percent of the irrigated land must be dedicated to drainage ponds. In California's San Joaquin valley, pond areas totaling 10–15 percent of the farmland area being drained are fairly typical, and this figure can rise to 50 percent when drainage systems capture a fair amount of salty groundwater as well.[30]

In addition, as happened in the Tulare basin in California, evaporation ponds can become an enticing but toxic habitat for migrating waterfowl, especially in areas where natural wetlands have largely disappeared. Moreover, these ponds offer only a short-term solution to a long-term problem. Eventually, the salt crust builds up and the pond becomes useless. The salted-out land must then either be decommissioned and cleaned up or sacrificed—both of which pose additional environmental problems.

While the search for lasting and cost-effective solutions to the drainage problem continues, several promising approaches could greatly reduce the volume of drainage needing disposal. The first line of attack is to irrigate more efficiently. In just about every area where salt is a problem, a wide variety of efficiency improvements—from scheduling irrigations properly to installing drip irrigation—could greatly reduce waterlogging and salinity problems. (See Chapter 8.) By applying only as much water as crops need during the growing season, plus just enough to leach away accumulated salts in the root zone, farmers can reduce waterlogging and drainage, and keep more land in production than if they persist with poor management practices.[31]

Along with more-efficient methods, farmers can capture and reuse drainage water to irrigate crops with a higher tolerance for salt. Much drainage water that is not considered "fresh" by conventional water quality standards may nonetheless be perfectly suitable for irrigating salt-tolerant crops. In the western Negev of Israel, for example, cotton is successfully grown by irrigation with highly salty water from a local saline aquifer. The Israelis have also found that certain crops—such as tomatoes grown for canning or pastes—may actually benefit from somewhat salty irrigation water. By directly reusing drainage water, irrigators can keep rivers cleaner, stretch regional water supplies, and avoid the sacrifice of irrigated land for evaporation ponds.[32]

James Rhoades and his colleagues at the U.S. Salinity Laboratory in Riverside, California, have pioneered this reuse approach and achieved promising results. In field trials under commercial farming conditions in California's Imperial Valley, they grew wheat, sugarbeets, and cantaloupes in a two-year rotation, and repeated that rotation once, for a four-year trial. Because most crops tend to be more sensitive to salt during their germination and seedling phases, the researchers used relatively higher quality Colorado River water for the preplanting and early irrigations of wheat and sugarbeets, but then used saltier drainage water for later irrigations. The melons, grown last in the sequence and sufficiently salt-tolerant to withstand the salty soil left by the two previous crops, received only the better quality water. Not only did this cleaner water aid the melons' growth, it removed some of the salt from the soil. In this way, the cycle could begin again, with the planting of wheat, without much deterioration of soil quality.[33]

Remarkably, none of the crops grown in the study yielded significantly less than those in control plots. A similar experiment in which researchers grew two crops of cotton, followed by one of wheat, followed by almost two years of continuous alfalfa, was also successful. In this rotation, scientists used salty drainage to irrigate the two cotton crops after the seedlings had been established, and used the cleaner river water for the wheat and alfalfa. While the second cotton crop had lower yields, as expected, the researchers were primarily testing whether wheat, a moderately salt-tolerant crop, would grow all right in the salinized soil left by the cotton crops if it was irrigated with higher-quality water—and they found that it did.[34]

A wide variety of cropping sequences is possible because crops vary so much in their salt tolerance. (See Table 5–2.) The best choice for any particular farmer will depend on soil type, climate, the quality of both the incoming irrigation water and the drainage, and the farmer's management practices. But better matching crops to soil and water salinity can be a key to keeping irrigated agriculture productive as salinity problems worsen. A salt-sensitive crop such as onions, for example, may experience an 80-percent drop in yield when grown in soils with salt levels that do not affect barley or cotton at all.[35]

Farmers and scientists in Australia, California, and elsewhere have also experimented with planting thirsty trees, such as eucalyptus, and deep-rooted crops, such as alfalfa and lupins, to lower the water table in regions threatened with salinization. As with the engineering approaches, however, the long-term efficacy of these biological options has yet to be proved. Studies on the west side of California's San Joaquin valley have shown that a plantation of eucalyptus trees was able to consume salty groundwater and thereby lower the water table from about half a meter below the soil surface down to 2.3 meters. This reduced the immediate risk of salt buildup. But after several years of irrigating the eucalyptus trees with

Table 5–2. Salt Tolerance of Selected Conventional Crops

Tolerance	Crop
Sensitive	Apricot, Bean, Carrot, Grapefruit, Orange, Onion, Peach, Strawberry
Moderately Sensitive	Alfalfa, Corn (grain) Cucumber, Lettuce, Potato, Rice, Sugarcane, Tomato
Moderately Tolerant	Barley (forage), Sorghum, Soybean, Wheat
Tolerant	Barley (grain), Cotton, Sugarbeet, Wheat (semidwarf)

SOURCE: E. V. Mass, "Crop Salt Tolerance," in K.K. Tanji, ed., *Agricultural Salinity Assessment and Management* (New York: American Society of Civil Engineers, 1990), as presented in F. Ghassemi, A.J. Jakeman, and H.A. Nix, *Salinisation of Land and Water Resources* (Sydney: University of New South Wales Press, 1995).

poor-quality drainage, the soil became so laden with salt and other substances that the trees were no longer able to extract moisture from it. For this crop-tree system to continue functioning, more fresh water would need to be applied to the tree plantation to leach away the accumulated salts, which could reduce the amount of irrigation water available to grow crops.[36]

A third option is for governments to require that the most problematic irrigated lands be removed from production. The U.S. National Research Council reports, for example, that without measures to better manage and dispose of agricultural drainage, some portion of the irrigated cropland on the west side of the San Joaquin valley may need to come out of irrigated production. This is undoubtedly the case in other major irrigated areas as well.[37]

If irrigators cannot completely avoid the double jeopardy of waterlogging and salt buildup, can salted-out land at least be reclaimed or used for some other productive purpose? Again, the answer seems to be a qualified yes. Salt-loving plants called halophytes can thrive in many salty environments that are completely unsuitable for even the most salt-tolerant of conventional crops. They have evolved special mechanisms for managing salt that allow many of them to be irrigated even with seawater. Varieties of *Atriplex*, a type of saltbush, for example, have specialized cells called salt bladders on their leaves for storing surplus salt. When the bladders are full, they burst, releasing the salt. Cells inside each leaf have special ways of managing any salt that the plant absorbs. Although halophytes will not necessarily restore salty soils to a degree that allows conventional crops to be grown again, they can make those soils economically productive by contributing oilseeds, fodder for livestock, and other products.[38]

More than 20 years of work in western Australia has demonstrated that planting saltbush can convert salty wasteland into profitable grazing land, in some cases netting $100 per hectare. Many saltbushes are perennial shrubs that remain green year-round, and are good forage crops in dry regions. Australian scientists are also promoting the use of saltbush for land reclamation in Victoria. In Pakistan, a deep-rooted, highly salt-tolerant perennial called kallar grass shows promise. A nitrogen-fixer, kallar grass grows well without fertilizer and is a useful forage crop, producing about 50 tons of biomass per hectare. In some areas its growth has restored soils to the point where a few varieties of barley and moderately salt-tolerant crops will grow again.[39]

The bottom line is that there is no simple fix to the salt problem. Solutions require taking inventories of how extensive a problem salt is, monitoring its spread, diagnosing its source, testing possible remedies, evaluating how they work, and then putting a long-term strategy into effect. Few if any regions have acted with this degree of diligence. Even when solutions are identified, they are typically expensive, difficult to fund, and not always successful.[40]

By growing crops more tolerant of salt in salty soils, farmers can lessen the damage salinity causes to crop production. But they typically cannot avoid yield reductions entirely, because salt usually exacts a yield penalty. A plant uses more energy to extract moisture from salty water than from fresh water, energy that would otherwise go into its growth. Although breeding or bioengineering plant strains with greater salt-tolerance can help, those plants will rarely yield as much as those grown in less salty environments. As salinization worsens and spreads to more areas, crop yields will inevitably fall.

No one knows how much land worldwide ultimately will come out of crop production—either because salinization ruins the soil, land must be dedicated to storing salt or toxic runoff, or land retirement is the only way to lower salt and toxic burdens to more manageable levels. Salinization could be a time bomb waiting to explode upon the agricultural scene, or it could remain a quiet but steadily increasing suppressor of crop yields. Either way, salt remains one of the gravest threats to irrigated agriculture and food security in a world that will be striving to feed 8–9 billion people within 50 years.[41]

6

WATER WARS I:
FARMS VERSUS
CITIES AND NATURE

═════════════

Why would the billionaire Bass brothers of Texas, enriched by real estate and oil deals, make a play for a parcel of farmland in the hot, dry American Southwest? The reason has little to do with the lettuce, tomatoes, and melons grown in California's sun-drenched Imperial Valley, but everything to do with what allows crops to grow there—water. In purchasing more than 16,000 hectares of valley farmland, the Bass brothers were simply acting on a tip from an old but timely adage: water flows uphill toward money.[1]

About a fifth of the Colorado River's annual flow goes to the Imperial Irrigation District (IID), which irrigates nearly 200,000 hectares of cropland. Thanks to a century-old deal with the federal government, IID gets this water for free. Farmers within the district pay only for the cost of delivering the water, about 1¢ per cubic meter. A few hundred kilometers to the west in Los Angeles, the Metropolitan Water District (MWD), which is the water wholesaler for about 16 million southern Californians, pays up to 16¢ per cubic meter for water that it sells to its customers for about 28¢ a cubic meter—28 times as much as the IID farmers pay.[2]

To astute investors, the math is compelling enough. But politics is weighing in as well. California has been using about 14 percent more Colorado River water than a 1922 interstate agreement entitles it to, and the U.S. government has put the state on notice that it must find a way to live within its allotted share. Any cutbacks would come out of urban supplies, since the Imperial Valley farmers have more senior water rights, and thus higher priority.[3]

Not long after buying their IID farmland in 1994, the Bass brothers began pushing the irrigation district, which actually owns the water rights, to strike a deal with San Diego, MWD's biggest customer. Three years later, in late 1997, the Basses hedged their bets and traded their $60-million investment in Imperial Valley farmland for $250 million worth of stock in the United States Filter corporation, the world's largest water treatment company. Meanwhile, IID did manage to strike a deal with San Diego. In 1998, the irrigation district agreed to transfer up to 246.8 million cubic meters of water a year to San Diego at an initial price of 20–27¢ per cubic meter. San Diego residents will benefit from lower costs and greater reliability of future supplies. IID will reap substantial profits, and if most of the water transferred results from increased efficiency and shifts to less thirsty crops, farmers will not necessarily need to take land out of production.[4]

Water grabs and power plays are legendary in the western United States. In the popular movie *Chinatown*, Hollywood capitalized on the drama of Los Angeles sucking farms dry in the Owens Valley. American writer and humorist Mark Twain captured the West's tension over water with his famous quip that "whiskey's for drinking, water's for fighting about." But as water becomes more scarce, the stakes are rising—not just in the western United States, but in many other parts of the world as well.[5]

For rapidly growing cities and industries, agriculture holds the last big pool of available water. Globally, irrigation accounts for about two thirds of all the water removed from rivers, lakes, and aquifers, and in many important agricultural regions, it claims 80 percent or more. As opportunities to expand water supplies dwindle, competition over existing supplies is mounting. How this competition plays out is about much more than whether rich investors get richer. It is about food security, social stability, the health of rural communities, the plight of the world's poor, and the ability of the aquatic environment to continue supporting a diversity of life.

Losing Out to Cities

On an average day in the developing world, about 150,000 people join the ranks of urban dwellers. Some are babies born to couples already living there. Others migrate in from the countryside, hoping for a better life. Most need shelter and a job. All need food and water.[6]

By 2025, nearly 5 billion people are expected to live in cities, about twice as many as in 1995. At that time, urbanites will represent a majority—59 percent—of the world's population, up from 46 percent in 1996. Mark Rosegrant and Claudia Ringler of the International Food Policy Research Institute in Washington, D.C., project that annual water demands by households and industries in developing countries will climb by 590 billion cubic meters between 1995 and 2020, and that the share of water going to these activities will more than double—from 13 percent of total water use to 27 percent.[7]

It is a fairly sure bet that a portion of these increased urban and industrial demands will be met by transfers of water out of agriculture. What is not known is how much water irrigators will transfer, whether they will transfer the water voluntarily, and how much crop production will decline as a result of the transfers. These are important questions. As Rosegrant and

Ringler conclude, the way the farm-to-city reallocation of water is managed "could determine the world's ability to feed itself."[8]

If farmers make little effort to save water by irrigating more efficiently or growing less-thirsty crops, transfers of water out of agriculture will cause crop production to fall. Yields will decline in farming areas that lose or sell some of their water. In some cases, farmers will take cropland out of production altogether. If, for example, half of the projected rise in urban and industrial demand by 2020 is met by shifting irrigation water to these users, and little improvement is made in irrigation efficiency, grain production could drop by some 300 million tons—about 1.5 times current global grain exports.

A quick reality check shows that the shift of water from farms to cities is already under way, and is likely to increase. In parts of north China, including areas around Beijing, reservoirs that had supplied irrigation water to farms are often now used almost exclusively to supply households and factories. Farmers in Daxing County, about 50 kilometers south of Beijing, for example, no longer receive irrigation water that used to be shipped in by canal from Beijing's reservoirs. In 1993, they told a *New York Times* reporter that it had been more than a decade since local farmers could flood a rice paddy.[9]

Nationwide, China has been urbanizing at a rapid rate. The number of cities has climbed from 130 in 1949 to more than 600 today. About half of them are already short of water, and there is increasing pressure to pull supplies away from agriculture to narrow the urban water deficits. A mid-1990s planning study by China's State Statistical Bureau and Ministry of Water Resources concluded that 40 percent of the projected demand gap in 2000 could be met by shifting water out of agriculture. Moreover, these gaps will widen over the next couple of decades. Eugene Linden notes in *Foreign Affairs* that "the great urban migration has only just begun in China, which is still more than 70 percent rural." The United Nations projects that more than half of China's people will live in cities by 2025.[10]

In China, as elsewhere, both politics and economics drive water's reallocation. A cubic meter of water used in China's industries generates more jobs and about 70 times more economic value than the same quantity used in agriculture. As supplies tighten, water will shift to where it is more highly valued. Moreover, because only a small fraction of urban and industrial wastewater is treated before being released back to the environment, a growing share of China's rivers and streams are becoming too polluted to use—worsening the water crunch. The Huai River in central China, for example, is so polluted that officials have banned farmers from using it to irrigate crops. Canadian geographer Vaclav Smil, a specialist on China's environment, has estimated that as much as one fifth of China's river water is too polluted for irrigation use, much less for drinking.[11]

Farmers in India face mounting competition over water as well. India will add some 340 million people to its cities between 1995 and 2025, more than the current populations of the United States and Canada combined. Reallocations are reportedly occurring to increase supplies for the cities of Madras, Coimbatore, and Tirupur and for a number of smaller towns. Tirupur, in the southern state of Tamil Nadu, suffers from a water deficit of some 22 million cubic meters a year and serious degradation of water quality, and has begun to import more water from outside the city. Many farmers within 35 kilometers of the city have abandoned farming and instead sell their groundwater to urban and industrial users.[12]

Rice farmers in parts of the Indonesian island of Java are losing water supplies to textile factories, even though Indonesian law gives agriculture higher priority for water. A study of one irrigated region in West Java found that factories often take more water than their permits allow and also take it directly out of irrigation canals, leaving too little for the farms. Some factories buy or rent rice fields from farmers just to get access to the irrigation water, but then leave some of the fields fallow. The factories have also polluted local water supplies, which has lowered rice yields and killed fish in local fish ponds. Lacking legally enforced rights to the water they have been accustomed to using, the farmers have little recourse. Researchers Ganjar Kurnia, Teten Avianto, and Bryon Bruns note that "many farmers, suffering from lost production and insecurity of water supplies, feel they have no choice but to sell their land."[13]

Growing demand in the megacities of Southeast Asia, including Bangkok, Jakarta, and Manila, is already partially being met by overpumping groundwater, and so pressure will intensify to shift water out of agriculture in these regions as well. In addition to booming populations, cities in these areas also face increased demands from rising affluence. In Malaysia, for example, the number of golf courses has tripled over the last decade to more than 150, and 100 more are planned. Together, Malaysia, Thailand, Indonesia, South Korea, and the Philippines maintain 550 golf courses, with another 530 already in the planning pipeline. Besides chewing up farms and forests, golf courses in these countries typically require irrigation at the same time crops do—during the dry season, when supplies are usually tight.[14]

Overall, cities in industrial countries will likely pull less water out of agriculture because their water demands are rising much more slowly. But in rapidly urbanizing, water-short areas, such as the western United States, the city-farm competition is heating up. Cities are buying water, water rights, or land that comes with water rights in parts of Arizona, California, Colorado, and elsewhere. Not surprisingly, the biggest trades so far have involved the Imperial Irrigation District in southern California, which is within striking distance of urban areas that are home to 16 million people, and still growing.

In addition to the recent deal with San Diego, IID agreed to a trade in 1989 with the Metropolitan Water District in Los Angeles. MWD agreed to invest in efficiency improvements within IID in exchange for the water those investments save. The trade will shift up to 106,000 acre-feet (130.8 million cubic meters) a year from farm to urban uses for 35 years. MWD benefits because the cost of the conserved water will be less than 10¢ per cubic meter, much lower than its best new-supply option. IID benefits from the cash payments and an upgraded irrigation network. And because the water traded is generated through conservation, no cropland needs to come out of production.[15]

Another MWD deal, however, does require farmers to take land out of irrigated production. In 1992, the urban water wholesaler entered into an agreement with the Palo Verde Irrigation District, located on the west side of the Colorado River between Parker and Imperial dams. The agreement called for Palo Verde farmers to fallow a portion of their cropland for two years and transfer the resulting water savings to MWD.[16]

Facing unstable crop prices, 63 farmers signed on, fallowing a total of 8,181 hectares. MWD paid the irrigators $3,064 for each hectare left unplanted and, in return, received a total of 228 million cubic meters of water—the equivalent of about 10 percent of MWD's yearly deliveries. The transferred water was stored in federal reservoirs on the lower Colorado River for use any time MWD desired before the year 2000. As in its deal with IID, MWD benefits by obtaining additional supplies at a lower cost. Palo Verde farmers benefited from more stable income. But because land was taken out of production, farm workers lost jobs.[17]

Water transfers often affect people not involved directly in the sale, which makes a full accounting of costs and benefits hard to achieve. But the costs to so-called third parties, who rarely have a place at the negotiating table, can be substantial. These costs can also be cumulative, affecting rural communities, employment, the tax base, and the environment. Because poorer farm laborers may be the ones to lose jobs, even economically efficient water trades may worsen inequities. Water trades can also damage downstream wetlands and lakes. IID's deals with both the MWD and San Diego, for example, could harm the inland Salton Sea, an important stopover for many species of migratory birds. (See Chapter 5.) Though polluted, IID's drainage is critical to sustaining the area and quality of the sea, which is already 25 percent saltier than the Pacific Ocean. As IID sends increasing amounts of its water to southern California cities, the sea will shrink and become even saltier.[18]

In sum, the limited evidence to date suggests that the impacts of water transfers are decidedly mixed, complex, and difficult to predict. Moreover, especially in Third World settings, irrigation water is often used for many activities other than farming, including household activities, home gardens, livestock, and fishing. In these cases, third-party impacts can be substantial, and without compensation to the losers, can worsen inequities and deepen rural poverty.[19]

So far, few countries have the institutions and incentives in place to steer water competition in both a productive and an equitable direction, and to compensate those negatively affected by water trades. In Chile, where government policies encourage water marketing, negative impacts seem to be minimal in part because farmers typically sell relatively small portions of their water rights to cities, while at the same time investing in more-efficient irrigation technologies and practices. This allows them to maintain their crop production levels even as some water shifts from farming to urban uses. (See Chapter 10 for more on water markets.)[20]

Without a doubt, cities will continue to siphon water away from agriculture. What is not known is how much water ultimately will be reallocated and how great an impact that reallocation will have on food production, on farmers, and on rural economies. Unless this competition is managed well, it could dampen food supplies in some areas, and make the rich richer and the poor poorer. And competition for water may force more rural dwellers to head for the cities—which, in a vicious circle, would intensify the problem.

Nature Stakes a Claim

Irrigated agriculture faces another major competitor for fresh water—the natural environment. In recent years, ecologists, environmentalists, and concerned citizens have sounded alarms about the decline of rivers, lakes, wetlands, and other freshwater ecosystems. Increasingly, these groups are calling not only for fewer new dams and river diversions, but also for returning to natural systems some of the water now going to human activities—including irrigation.

During just this decade, public values have changed markedly in favor of protecting natural ecosystems and the multitude of benefits they offer, especially in wealthier countries. Fishing, kayaking, rafting, and other recreational pleasures top most people's interests in healthy rivers, but there is also greater awareness of the so-called ecosystem services provided by intact freshwater systems. These services include controlling floods, purifying water supplies, maintaining fish and wildlife habitat, and conserving species richness.

As more scientists, citizens, and political leaders speak out about the need to protect these functions of rivers and natural ecosystems, the balance of power governing water use is changing in ways that could revolutionize water management. Engineers built the hundreds of thousands of dams that now block the world's rivers with four principal goals in mind—flood control, hydroelectric power production, water supply, and irrigation. Almost without exception, they paid little attention to the downstream effects of altering river environments. Now that scientists have uncovered serious damage from dams, levees, and other hydraulic infrastructure, and the public has spoken out about its increasing environmental concerns, the water equation is shifting. Legislatures, courts, and the public increasingly view the natural environment as having a legitimate claim to water, and they are deciding to return some water to nature.

Dams, diversions, dikes, and levees destroy aquatic habitat. They sever the connections a river has with its floodplain, its channel, its delta, and the sea into which it empties. They change the temperature of a river's water and its pattern of flow throughout the year. They also prevent a river from performing most of its natural functions, such as delivering nutrients to the seas to sustain fisheries, absorbing floodwaters by spreading them over its floodplain, protecting wetlands and their ability to filter pollutants, providing habitat for a rich diversity of aquatic life, maintaining salt and sediment balances, offering myriad recreational opportunities, and providing some of the most inspirational natural beauty on the planet.

Scientists have just begun to uncover the extent and magnitude of damage to the world's freshwater environment, so it is impossible to know how much corrective action—and water reallocation—ultimately will be needed. The evidence to date, however, suggests that it will not be trivial. Swedish scientists Mats Dynesius and Christer Nilsson examined the 139 largest river systems in the United States, Canada, Europe, and the former Soviet Union—essentially the northern third of the world—and found that 77 percent of them were moderately to strongly altered by dams, reservoirs, diversions, and irrigation. The remaining 23 percent were relatively small systems located in the far north, mainly in the boreal and arctic regions. Writing in the journal *Science*, they warned that because of the extent of river exploitation, key habitats such as waterfalls, rapids, and floodplain wetlands could disappear entirely from some regions, extinguishing numerous plant and animal species specific to running waters.[21]

More detailed studies of species imperilment confirm that these risks are real. Worldwide, one out of every three fish species is threatened with extinction, compared with one out of every four mammals, one out of every five reptiles, and one out of every nine birds. In the United States, the most striking finding of the most comprehensive assessment to date on native plant and animal species is the dire condition of species that depend on aquatic systems for all or part of their life cycle. Conducted by The Nature Conservancy and the Natural Heritage Network, the 1997 study found that 67 percent of freshwater mussels are at risk of extinction, along with 51 percent of crayfish, 40 percent of amphibians, and 37 percent of freshwater fish. As a whole, freshwater species are more in jeopardy than land-based species, and the leading cause of their imperilment is the destruction and degradation of their habitats.[22]

The legal and institutional means for protecting freshwater habitats vary by country. But even in the United States, conservation groups and others concerned about species loss face an uphill battle. Historically, the U.S. Congress gave control over water rights and allocations to the states, and there was little federal intervention. With passage of the Wilderness Act, the Clean Water Act, the Wild and Scenic Rivers Act, the National Environmental Policy Act, the Endangered Species Act (ESA), and other federal legislation, tensions arose over state-granted private water rights (which, under western water law, are as firm as property rights) and the new federal laws that were aimed at protecting the broader public interest. Although Congress has generally deferred to state water laws, the Supremacy Clause of the U.S. Constitution says that when conflicts between federal and state laws arise, federal law prevails. This means, for example, that state-granted water rights for irrigation may have to bend to requirements to protect critical habitat for species listed as endangered under the federal ESA.

For irrigators in the western United States, this possibility is no small concern. Of 68 fish species listed as threatened or endangered in the 17 western states in 1993, 50 had "agricultural activities" recorded as one of the factors behind their decline. Nearly one out of every five western counties contains habitat for one or more of these 50 species. Irrigated areas in California, Colorado, Idaho, and Utah that rely extensively on river water correspond closely with areas harboring high concentrations of ESA-listed species.[23]

Moreover, the number of listed species continues to grow. Since 1967, when 12 western fish species were found to be endangered under a law that preceded the ESA, the number of threatened or endangered western fish species has risen nearly sixfold. A species remains on the list until either recovery efforts sufficiently diminish threats to its survival or it becomes extinct.[24]

Federally built dams and diversions supply more than a third of the surface water consumed by irrigated agriculture in the West, and many of them are implicated in species destruction. They threaten Chinook salmon in the Pacific Northwest's Columbia River basin and in California's Sacramento basin, two fish species in Nevada's Truckee River–Pyramid Lake system, and the Colorado River squawfish in the Colorado basin, to name a few. Because the ESA requires federal agencies— including the Bureau of Reclamation, the largest water supplier to western irrigators—to ensure that their actions are unlikely to jeopardize a listed species, federal projects are in the front line of battles between species protection and traditional water uses. As University of Colorado law professor Charles Wilkinson puts it, the ESA could "prove to be a sturdy hammer for dislodging long-established extractive water uses that have worked over so many western watersheds and drained them of much of their vitality."[25]

That hammer has already been struck in a number of river basins. In late 1992, the U.S. Congress passed legislation aimed at revamping operation of the huge federal Central Valley Project in California in order to shore up the health of the Sacramento–San Joaquin River system. Among other aims, the law set a goal of restoring the natural production of salmon and other anadromous fish (those that migrate from salt water to fresh water to spawn) to twice their average levels over the preceding 25 years.

The Sacramento River has four salmon runs, each designated by the time of year the fish pass under the Golden Gate Bridge to begin migrating upstream. The population of winter-run chinook plummeted from a peak of 117,808 in 1969 to 533 in 1989—the year authorities listed the fish under the ESA. The 1992 law dedicates 800,000 acre-feet (987 million cubic meters)—about 10 percent of the Central Valley Project's annual water supply—to maintaining fish and wildlife habitat.[26]

Irrigators, not surprisingly, protested the loss of their water and fought relentlessly to roll back the reforms. After five years of haggling, the Department of Interior issued a compromise proposal in late 1997 that called for varying water allocations for fish according to river and fish conditions rather than abiding by the law's fixed allocation. Environmental groups fought back, and in April 1998, the Department of Interior issued a policy that ensures adequate water deliveries to wetlands and wildlife refuges in the Sacramento–San Joaquin River basin.[27]

California's water battlegrounds also include the San Francisco Bay delta, a highly productive aquatic ecosystem that harbors more than 120 species of fish (including the endangered delta smelt and the winter-run chinook salmon) and supports 80 percent of the state's commercial fisheries. The delta supplies water to some 20 million Californians, as well as for irrigating 45 percent of the nation's fruits and vegetables. For many years, fierce conflict has raged over how to balance competing demands on this hub of California's water system.[28]

Protracted negotiations among federal and state officials, farmers, environmentalists, and other affected parties led to a much-heralded consensus agreement in 1994. That accord laid out short-term water quality and outflow requirements for the delta and called for a three-year effort to develop a long-term solution. Five years later, the quest to forge a consensus-based solution continues, with attention and public comment now focused on a revised plan issued in December 1998. A final "preferred alternative" is expected to be announced by the end of 1999.[29]

In neighboring Nevada, home of the nation's first major federal irrigation project, agriculture has clearly lost some rounds in the shifting balance of power over water. Since 1903, irrigation projects have siphoned off much of the flow of the Truckee and Carson Rivers, the lifeblood of northern Nevada. Named after Senator Francis Newlands, architect of the 1902 Reclamation Act, the Newlands irrigation project diverted the Truckee, which flows out of Lake Tahoe down the eastern slope of the Sierra Nevada, past Reno, across 50 kilometers of desert and then through 800 kilometers of canals.

Over time, wetlands and lakes that had been sustained by these rivers began to dry up. Winnemucca Lake, once a wildlife refuge, disappeared in 1938. The Stillwater wetlands declined and were fed increasingly by toxic drainage running off of farm fields. Pyramid Lake, the Truckee's final destination, began to shrink from the diminished inflow, bringing two native fish species to the brink of extinction—the Lahonton cutthroat trout and the cui-ui. The shrinking fish populations alarmed the Paiute Indians, whose reservation surrounds the lake, since they depend on cutthroat trout anglers for income and view the cui-ui, a relic of the Ice Age, as central to their social, religious, and culinary traditions. In fact, the Paiutes' ancient name, Kuyuidokado, means "the cui-ui eaters."[30]

A coalition of federal, state, Indian, and environmental interests have now reordered the priorities for water use in the basin. Not surprisingly, the biggest losers will be the century's largest water consumers—irrigated farms. The several thousand farmers in the Newlands project used to receive more than half of the Truckee River's flow, but their share has now dropped to about a fifth. "For about 100 years it was everybody against the Indians," a lawyer for the Paiute told the *New York Times* in 1997. "In a very short period of time, that's turned around. Now, it's everybody against the Newlands project." With revolving loan money available through the federal Clean Water Act and funds from the Department of Interior, officials are now buying irrigation water rights and restoring flows to the Truckee River.[31]

In addition to public and private efforts to reallocate water to the environment, dams—the ultimate symbol of human control over water—are coming under greater scrutiny. A few dams have already been slated for removal because officials have judged their environmental damages, which were long overlooked, to outweigh their benefits. For example, officials have called for the removal of several hydropower dams in order to restore fisheries and recreational opportunities. Among them are Edwards Dam on Maine's Kennebec River and the Elwha and Glines Canyon dams on the Elwha River in Washington state.[32]

Much bigger proposals are afoot as well that collectively are nothing short of revolutionary. The U.S. Army Corps of Engineers is now studying the idea of breaching four dams on the Lower Snake River in the Pacific Northwest. Besides blocking the river, the dams create continuous slack water for about 220 kilometers upstream of the Snake's confluence with the Columbia, causing most salmon and steelhead to die during their migrations. The *Idaho Statesman*, a conservative newspaper in a conservative state, has endorsed the proposal.[33]

As one species after another has been added to the endangered species list, the dam debate has deepened and divided western interests. A dozen stocks of West Coast salmon, steelhead, and trout are now listed as threatened or endangered, and an additional 13 have been proposed for listing. Across the United States, some 200 separate runs of fish have gone extinct. Upon signing the landmark agreement clearing the way for the removal of Edwards Dam in Maine, Secretary of Interior Bruce Babbitt said that this "is a challenge to dam owners and operators to defend themselves—to demonstrate by hard facts, not by sentiment or myth, that the continued operation of a dam is in the public interest, economically and environmentally."[34]

Over the last two years, the radical idea of breaching one of the Colorado River's megadams—Glen Canyon—and draining massive Lake Powell has gained force. The proposal has even gotten a hearing on Capitol Hill. Daniel P. Beard, former Commissioner of the reclamation agency that built Glen Canyon Dam, wrote in a 1997 *New York Times* editorial: "There is greater competition for water between cities and farms. Federal construction money has dried up, and environmental concerns have become more urgent. Draining a reservoir and restoring a canyon may just be the cheapest and easiest solution to our river restoration problems."[35]

Although they lack the drama of dam deconstruction, proposals to operate dams according to reordered priorities may have an even larger effect. The driving idea is to manage dams so as to recreate the river's natural flow patterns, thereby benefiting native species and the river system as a whole. Flaming Gorge Dam on the Green River in Utah, for example, is now operated in this way. After the U.S. Fish and Wildlife Service invoked the Endangered Species Act to protect critical habitat for endangered chubs, suckers, and squawfish, officials began dictating dam operations not by irrigation, flood control, and power needs, but in a way that would recreate natural habitat.[36]

Generally, this approach has involved trying to recreate the pre-dam flow patterns of the Green, a major tributary of the Colorado. Rather than storing as much water as possible for the peak irrigation and hydropower demands of late spring and summer, dam operators release a surge of water in May in order to mimic the natural spring flood and facilitate the spawning of native fish populations. Flows are then gradually reduced to a much lower level during the summer, simulating pre-dam conditions.[37]

Further downstream on the Colorado mainstem, federal officials have put similar regulations in place for Glen Canyon Dam in order to partially restore the natural habitat of the famously beautiful Grand Canyon. On the heels of promising results from a landmark test flood in March 1996, dam operators will likely release a major flood surge every 7 to 10 years, and smaller ones every spring. (Should the dam actually be breached, the regulations would of course become moot.)[38]

Although water rights and environmental mandates differ from one country to another, the conflict between private users and the public's interest in ecosystem protection is playing out elsewhere as well. In Australia, concern about the deteriorating health of the Murray-Darling River system is forcing an overhaul of water use and management there. Australia's largest river system, the Murray-Darling supplies about three fourths of all the water used nationwide. It is the main source of water for 16 cities, including Adelaide and Canberra, and some 70 percent of Australia's irrigated agriculture occurs within the basin.[39]

Together, farms, cities, and industries drain off 75–80 percent of the river's annual flow, leaving little for fish and other instream needs, especially during dry periods. The Murray-Darling harbors some 29 indigenous fish species, and—as in the United States—dams and river diversions are destroying their habitats. No national law akin to the U.S. Endangered Species Act safeguards these fish. A few Australian states have legislated the listing of threatened species, but have not followed up with management plans. In New South Wales, fish are not eligible for listing under threatened species legislation. Queensland has no such legislation at all.[40]

Nonetheless, the Murray-Darling basin states have recognized the dire condition of the river system, and have agreed to allocate 25 percent of the river's natural flow to maintaining the system's ecological health. This action is an important step forward, but its implementation—including negotiating each state's respective quota—will not be easy. In 1995, a basinwide freeze was placed on withdrawals for irrigation. In 1997, the Murray-Darling Basin Commission recommended capping allocations to major cities and towns at projected year-2000 levels of water use. The Commission is suggesting that cities meet any demands above this level by purchasing water from irrigators. Thus in Australia, as in the western United States, farmers practicing irrigated agriculture will either learn to make do with less water or take land out of production.[41]

In developing countries, where demands for food and water are rising rapidly, environmental claims generally have a lower priority and compete for water less successfully. But at least a few regions are considering reallocating water back to nature as the cost of ecosystem deterioration becomes more apparent. The five principal countries of the Aral Sea basin, for example, have agreed that the sea itself should be regarded as an independent water user, and that the ecosystem deserves an allocation of river water just as the countries do. Conceptually, this is a huge and important breakthrough, but implementing the idea is extremely difficult.[42]

Any serious restoration of the Aral Sea would require a massive shift of water out of irrigated agriculture. Nikita Glazovsky, Deputy Director of the Institute of Geography of the Russian Academy of Sciences, estimates that stabilizing the sea at roughly its 1990 level would require that 35 billion cubic meters of river water flow into it each year—about five times the average annual inflow registered during the 1980s. To free up this much water by retiring farmland, more than half of the basin's irrigated cropland would have to come out of production—an unthinkable scenario given the region's dire economic and employment conditions. Discussions have generally focused on more modest restoration of the sea and the river deltas. Such efforts might require on the order of 13–19 billion cubic meters a year of river inflow to the sea—still a large portion of agriculture's current supply.[43]

As discussed in later chapters, farmers in the Aral Sea basin could save a great deal of water by improving irrigation efficiency and shifting cropping patterns. Such steps would also lessen the basin's terrible salt problem. But the incentives and institutions needed to accomplish these transitions on a large scale do not yet exist.

Political leaders in the Aral Sea basin continue to hang their hopes on a large diversion of water from Russia's Siberian rivers into the Aral Sea basin—a controversial, decades-old, $40-billion proposal that President Mikhail Gorbachev shelved in 1986 because of its high price tag and environmental risks. The scheme is a classic example of the tendency to look to unrealistic engineering solutions to water problems in order to avoid the politically difficult but more lasting solution of adjusting economic and agricultural activity to the limits of the available water supply. Meanwhile, even though political leaders have agreed on the need to shift water back to the environment, four out of the five basin countries have plans to expand irrigation to new lands. Such action would set the stage for even more intense competition between agriculture and the environment in the years ahead.[44]

Water Stress and the Global Grain Trade

Will the growing competition for water have global effects, besides local, regional, and national ones? The answer depends in part on how water is transferred from farms to cities and from farms back to the environment. There is, however, one fairly certain global impact that few researchers and political leaders have noticed—the effect regional water scarcity and competition will have on the global grain trade.

Because water is so unwieldy and expensive to transport long distances, countries running short rarely import it. Instead, they import grain—which Tony Allan of the University of London has called "virtual water." With each ton of grain representing about 1,000 tons of water, countries in effect balance their water books by purchasing grain from other countries rather than growing it themselves.[45]

Most economists view this practice as a sensible way to respond to water shortages. They point out that water-scarce countries can generate much more income from their limited water by using it in commercial and industrial enterprises and then purchasing their grain on the international market. Israel, for example, has done nicely with this approach. As long as surplus food is produced elsewhere, nations with surpluses are willing to trade, and the countries that need food can afford to pay for the imports, it would seem that water-short countries can have food security without needing to be food self-sufficient.

This tidy logic is shaken, however, by the rapidly growing number of people who will be living in countries forced by water scarcity to follow this path. As a nation's net precipitation (also called runoff) per person drops below about 1,700 cubic meters, food self-sufficiency becomes difficult, if not impossible. In most countries, it is only possible to store and control 20–50 percent of the total runoff, so only a fraction of the water resource is actually available for use. As a result, below 1,700 cubic meters per person, there is often not enough usable water to meet the demands of industries and cities and to grow enough food for the entire population while at the same time sustaining river flows for navigation, fisheries, and other ecological functions. Countries in this situation then begin to import water indirectly, in the form of grain.

At present, 34 countries in Africa, Asia, and the Middle East have per capita runoff levels below 1,700 cubic meters a year. All but two of them—South Africa and Syria—are net importers of grain, and 24 of them already import at least 20 percent of their grain. (See Table 6–1). Collectively, these water-stressed countries import nearly 50 million tons of grain a year. World grain exports total about 200 million tons a year, so water scarcity is to some degree driving about one fourth of the global grain trade.[46]

As populations grow, per capita runoff will drop below the 1,700-cubic-meter level in more and more countries, and the countries already in this group will also have larger populations. By 2025, Africa will add 10 countries to this list. India,

Table 6–1. Grain Import Dependence of Selected Countries in Africa, Asia, and the Middle East with Less than 1,700 Cubic Meters of Annual Runoff per Person[1]

Country	Internal Runoff per Capita, 1995	Net Grain Imports as Share of Consumption[2]
	(cubic meters per year)	(percent)
Jordan	249	91
Israel	309	87
Libya	115	85
South Korea	1,473	77
Algeria	489	70
Yemen	189	66
Tunisia	393	55
Saudi Arabia	119	50
Uzbekistan	418	42
Egypt	29	40
Azerbaijan	1,066	34
Turkmenistan	251	27
Morocco	1,027	26
Somalia	645	26
Rwanda	808	20
Iraq	1,650	19
Kenya	714	15
Sudan	1,246	4
Burkina Faso	1,683	2
Burundi	563	2
Zimbabwe	1,248	2
Niger	380	1
South Africa	1,030	–3
Syria	517	–4

[1]Ten other countries have fewer than 4 million people each and are omitted from this table. Runoff figures do not include river inflow from other countries, in part to avoid double-counting. [2]Ratio of annual net grain imports to grain consumption averaged over the period 1994–96.
SOURCE: See endnote 46.

Pakistan, and several other Asian nations will join it as well. The total number of people living in African, Asian, and Middle Eastern countries with per capita runoff below the benchmark level will jump more than sixfold by 2025—from about 470 million to more than 3 billion. (See Table 6–2.) The vast majority of these people will be living in Africa and South Asia, where the deepest pockets of poverty and hunger are today.[47]

Like an M.C. Escher drawing, this larger picture of water scarcity's implications only comes into focus by standing back and absorbing all the parts at the same time. What appears to be a solid and sensible recommendation for any one country may appear just the opposite when applied to many. It seems dangerous to presume, as many economists and officials do, that there will be enough exportable grain to meet the import needs of all these countries at a price they can afford. And with world food aid at its lowest level since the mid-1950s, having dropped two thirds since 1992–93, relying on the generosity of grain-surplus nations to fill food gaps is a risky strategy.[48]

Water, long left out of the food security equation, may now

Table 6–2. Number of People in African, Asian, and Middle Eastern Countries with Less than 1,700 Cubic Meters of Annual Runoff per Person, 1995, with Projections for 2025

Region	1995	2025	Factor Increase
	(million people)		
Africa	295	908	3.1
Asia	86	1,957	22.8
Middle East	86	185	2.1
Total	467	3,050	6.5

SOURCE: See endnote 47.

be driving it. As domestic competition for water spills over into international competition for grain, it will be the poor, food-deficit nations that lose out. Without a concomitant rise in the income levels of the very poor, a rise in food prices could place the health and lives of many additional millions at risk. Confronting this threat head-on will take efforts to raise the food production and income levels of the poor directly. And as described in Chapter 9, irrigation has a key role to play in meeting this challenge.

7

WATER WARS II: IRRIGATION AND THE POLITICS OF SCARCITY

The nations of the region will act rationally
once they've run out of all other possibilities.

Abba Eban

Housed in the Louvre in Paris is what is thought to be the world's oldest recorded treaty. It was crafted around 3100 B.C. by two ancient Mesopotamian city states, Lagash and Umma, after a bloody battle over a water dispute. Lagash won the skirmish, killing Umma's governor in the process, and a boundary canal was dug and filled with Euphrates river water. But the treaty did not hold. Umma's army waged war against Lagash, killing its governor in return and perpetuating a 150-year conflict over territorial borders and access to water.[1]

Battles over water, and the use of water as a weapon in battle, checker the last 5,000 years of human history. Ancient legends, myths, Biblical accounts, and historical records offer fascinating accounts of water's role in political and military events. Destroying an enemy's irrigation networks, cutting off water supplies, and damming or diverting rivers to purposefully flood enemy territory are among the more common acts of water warfare carried out since ancient times.[2]

Today, threats of water wars persist. Shortly after signing

the historic peace accords with Israel in 1979, Egyptian President Anwar Sadat proclaimed that "the only matter that could take Egypt to war again is water." About a decade later, Egypt's foreign minister (and later U.N. Secretary General) Boutros Boutros-Ghali echoed the point, predicting that "the next war in our region will be over the waters of the Nile, not politics." King Hussein of Jordan declared in 1990 that water was the only issue that could take his country to war with Israel. And in 1995, World Bank vice president Ismail Serageldin proclaimed that "the wars of the next century will be over water."[3]

But will they? The highest purpose in making a gloomy prediction is to focus attention on properly diagnosing a problem so that society can act to ensure that the prediction turns out to be wrong. There is little doubt that a new politics of water scarcity is taking shape that threatens the security of nations and the stability of civil societies. Water wars, however, are not foreordained. Water scarcity provides more reason for neighboring countries to cooperate than to fight. Political leaders, water strategists, and diplomats face the challenge of ensuring that, at critical junctures, nations choose the rational course.

Anatomy of Water Disputes

The first step, arriving at a clear diagnosis, is by no means simple. On the one hand, a look for international incidences in which governments mobilized armies or when soldiers fired shots over water turns up surprisingly few cases. Aaron Wolf, an authority on water politics and a professor at Oregon State University, systematically searched for twentieth-century armed conflicts that qualified as "international crises" and that were at least partially caused by disagreements over water. He found only seven, and shots were actually fired in just four of these. (See Table 7–1.)

As Wolf is quick to point out, however, instances of water

conflict at the local or state level—for example, between tribes, communities, different water-using groups, or states within a country—are quite prevalent. Moreover, even if shots are not fired outfight, tensions over water can worsen relations between countries, making other issues harder to resolve and regional stability less secure. Perhaps most important, the small number of international armed conflicts over water during the twentieth century may not be a good predictor of possibilities for conflict in the twenty-first century.

First, water problems have worsened rapidly in just the last decade or two in many international river basins, and they continue to worsen. Second, ethnic tensions and social inequities have fueled a great deal of water-related violence within countries. With the breakup of nations and political blocks since World War II, the stage could be set for more domestic violence and cross-border conflicts in the next century. Africa, the Middle East, and the former Soviet bloc come first to mind. But political dissolution in large, ethnically diverse nations such as China, India, and Indonesia could spawn resource conflicts as well.

Three interacting forces drive tension and conflict over a resource such as water: the depletion or degradation of the resource, which shrinks the "resource pie"; population growth, which forces the pie to be divided into smaller slices; and unequal distribution or access to the resource, which means that some get larger slices than others. These forces can act either within countries or between them. Even conflict within countries can have destabilizing international repercussions. It can cause a regime to become more authoritarian or militant, lead to migrations of environmental refugees, or cause famine or other humanitarian disasters. In addition, water's special properties make it a strong candidate for inciting tensions between countries. Thomas Homer-Dixon of the University of Toronto, who has pioneered research into the links between resource scarcity and violent conflict, maintains that "the

Table 7–1. International Conflicts Linked to Water Disputes, Twentieth Century[1]

Countries/Basin	Years	Description
India and Pakistan; Indus River Basin	1948	Partitioning of British India awkwardly divides Indus tributaries and irrigation system; disputes over irrigation water bring the two countries to the brink of war.
Israel and Syria; Jordan River Basin	1951–53	Israel and Syria sporadically exchange fire over Israel's water development activities in the demilitarized zone between the two countries. Israel ends up moving its water intake to the Sea of Galilee.
Egypt and Sudan; Nile River Basin	1958	Pending negotiations over the Nile River, Egypt sends a military expedition into disputed territory between the two nations. Tensions ease with a change in Sudan's government and the signing of a Nile waters treaty.
Ethiopia and Somalia	1963–64	Border skirmishes occur over disputed territory in Ogaden desert that includes critical water sources; several hundred killed before a cease-fire is reached.

renewable resource most likely to stimulate interstate war is river water."[4]

Unique among strategic resources, water flows naturally across political boundaries. Many countries rely on river water arriving from upstream neighbors for a substantial portion of their surface supplies. (See Table 7–2.) Egypt, Iraq, Syria, Turkmenistan, and Uzbekistan, for example, each depend on rivers flowing through neighboring countries for two thirds or more of their total surface water. Particularly in the face of rising

Table 7–1 *(continued)*

Countries/Basin	Years	Description
Israel and Syria; Jordan River Basin	1965–66	The two countries exchange fire over an Arab plan to divert the head-waters of the Jordan River, presumably to frustrate Israeli efforts to build its National Water Carrier; Syria halts the diversion effort in 1966.
Iraq and Syria; Euphrates River Basin	1975	Iraq claims that the Euphrates flow reaching its border is intolerably low; upstream Syria claims that less than half the river's normal flow is reaching its border with Turkey; Syria closes its airspace to Iraq; both sides reportedly send troops to their common border; mediation by Saudi Arabia unlocks the tension.
Mauritania and Senegal; Senegal River Basin	1989–91	Ethnic tensions and inequitable access to the Senegal River flood-plain cause civilians from border towns to attack each other, killing several hundred. Governments of both countries use their armies to end the violence.

[1]Includes only cases in which armed conflict occurred or the probability of military hostilities was high; many more less acute water-related disputes have occurred.

SOURCE: Adapted from Aaron T. Wolf, "Conflict and Cooperation Along International Waterways," *Water Policy*, vol. 1, pp. 251–65 (1998).

water demands from population growth and irrigation expansion, such countries are highly vulnerable to decisions by upstream nations to siphon off more water for their own use.

Although this upstream-downstream tension has existed for centuries, the stakes are higher than ever. In a growing number of shared river basins, there is simply not enough

Table 7–2. Dependence on Imported Surface Water, Selected Countries

Country	Share of Total Flow Originating Outside of Border
	(percent)
Turkmenistan	98
Egypt	97
Hungary	95
Mauritania	95
Botswana	94
Bulgaria	91
Uzbekistan	91
Netherlands	89
Gambia	86
Cambodia	82
Syria	79
Sudan	77
Niger	68
Iraq	66
Bangladesh	42
Thailand	39
Jordan	36
Senegal	34
Israel[1]	21

[1]Includes only flows originating outside current borders; a significant additional share of Israel's fresh water originates from occupied, disputed territories.
SOURCE: Turkmenistan and Uzbekistan figures from David R. Smith, "Climate Change, Water Supply, and Conflict in the Aral Sea Basin," paper presented at the "Pri-Aral Workshop 1994," San Diego State University, March 1994; others from Peter H. Gleick, *Water in Crisis* (New York, Oxford University Press, 1993).

water to meet all projected needs. Irrigated agriculture is driving much of the competition since it accounts for 70–90 percent of water use in many of these regions. Prospects for the sustainability and future expansion of irrigation thus hinge, in part, on how nations resolve their water disputes.

Hostilities are most likely to erupt when the volume of water in a river is insufficient to meet all projected demands, when water use or allocation among the basin countries is perceived as inequitable, and when there is no recognized water-sharing treaty that includes all the countries in a river basin. These three conditions now exist in a number of locations. Moreover, rapid population growth guarantees that water tensions in several of them will worsen markedly. In five of the world's hot spots of water dispute—the Aral Sea region, the Ganges, the Jordan, the Nile, and the Tigris-Euphrates—the population of the nations within each basin is projected to climb between 44 and 75 percent by 2025. (See Table 7–3.) These countries will face heightened internal stress over water as agriculture and cities compete for limited supplies. They also confront hard choices about how to handle mounting tensions with neighboring countries.

No Longer Deep and Wide

The Jordan basin is by far the most water-short region. All three of the core parties in the basin—Israel, Jordan, and Palestine (the West Bank and Gaza)—are running up water deficits to meet current demands. Israel's deficit is projected to at least triple, from 200 million cubic meters (mcm) in 1991 to 600 mcm in 2020. Over the same time, Jordan's deficit is expected to increase four- to sevenfold. Altogether, Israel, Jordan, and Palestine will be running a water deficit of 1–2 billion cubic meters (bcm) a year by 2020, depending on total migration into the region and levels of water use per person.[5]

Such projections obviously cannot materialize. In some manner, demands will adjust downward to match the available supply. The main opportunities for narrowing the projected gaps lie in slowing population growth, using water more efficiently, and shifting water out of agriculture. Both Israel and Jordan have given up hopes of food self-sufficiency and rely on

Table 7–3. Populations in Selected Hot Spots of Water Dispute, 1997, with Projections to 2025

River Basin/Countries	Total 1997 Population	Projected 2025 Population	Change
	(million)		(percent)
Aral Sea[1] Kazakhstan, Kyrgyzstan, Tajikistan, Turkmenistan, Uzbekistan	55	83	+ 51
Ganges Bangladesh, India, Nepal	1,136	1,646	+ 45
Jordan Gaza, Israel, Jordan, Lebanon, Syria, West Bank	32	57	+ 78
Nile Burundi, Dem. Rep. of Congo, Egypt, Eritrea, Ethiopia, Kenya, Rwanda, Sudan, Tanzania, Uganda	299	497	+ 66
Tigris-Euphrates Iraq, Syria, Turkey	103	156	+ 51

[1]Excluding Afghanistan and Iran, which hydrologically are part of the basin.
SOURCE: Global Water Policy Project; population projections from Population Reference Bureau, *1998 Population Data Sheet*, wallchart (Washington, DC: 1998).

international markets for a substantial—and increasing—portion of their food. Syria, which receives more rainfall and substantial water from the Euphrates River, remains self-sufficient in grain. But with a population projected to climb 70 percent by 2025, that position may be difficult to maintain.[6]

Despite its acute water scarcity, the Jordan basin is unlikely to erupt in interstate conflict over water. Through military action during previous wars, Israel has succeeded in greatly improving its hydrologic position in the basin, and no other basin state will likely challenge Israel in the foreseeable future. In addition, the parties have taken some positive steps toward settling their differences. The Israeli-Jordanian peace treaty signed in October 1994 included an Israeli commitment to provide an additional 50 mcm a year to its neighbor. After Jordan complained about Israel's slow progress in living up to this promise, the two nations agreed in May 1997 that Israel would immediately begin sending 30 mcm a year to Jordan from the Sea of Galilee, with the remaining 20 mcm to come from a desalination plant to be built over the next three years.[7]

With the September 1995 signing of the interim agreement between Israel and the Palestinians, Israel recognized for the first time that the Palestinians have legitimate rights to West Bank water. The actual amount of those rights was left to be determined during the "final-status" talks. These are positive developments, but with the peace process still mired down as of early 1999, it remains to be seen whether progress continues on these bilateral water fronts. Ultimately, water security depends on all five parties in the basin arriving at an equitable sharing of the basin's waters—and such an agreement appears to be a long way off.[8]

Who Will Irrigate with the Nile?

The picture in the Nile basin is in some ways more unsettling. The basin has 10 nations within it, but the two most downstream—Egypt and Sudan—have staked claims to the vast majority of the river's flow. (See Figure 7–1.) By virtue of a 1959 treaty between the two countries, Egypt claims rights to 55.5 billion cubic meters a year and Sudan to 18.5 bcm. Egypt is already bumping up against the limits of its allotment, even as its population of 66 million leaps by an additional 1 million every eight months. Egypt has plans to expand its irrigated area by at least 1 million hectares over the next 20 years, which could easily require an additional 8 bcm of water. This planned

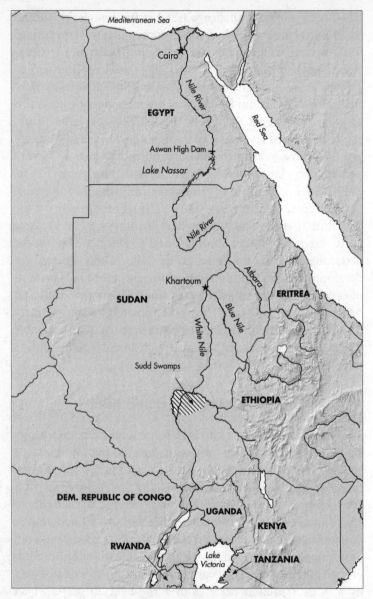

Figure 7–1. The Nile Basin

expansion sets the stage for conflicts with upstream countries that would like to begin siphoning off more Nile water for their own development—especially Ethiopia, where 86 percent of the Nile's flow originates.[9]

After years of civil war and social upheaval, Ethiopia now has the political stability to begin tapping upper Nile waters for its own agricultural and economic advancement. Some 3.7 million hectares of Ethiopia's land could potentially be irrigated, but only 5 percent of this area currently receives irrigation water. In recent years, Ethiopia has dusted off and updated a 1964 plan devised by the U.S. Bureau of Reclamation that had proposed 33 irrigation and hydropower projects in the upper Blue Nile region, and that would bring irrigation water to 434,000 hectares of land.[10]

The plan also called for four major hydroelectric dams on the Blue Nile, with a combined reservoir storage capacity of about 50 bcm a year—nearly equal to one year's total flow of the Blue Nile. The creation of new reservoirs—whether for hydropower or water supply—will reduce flows downstream, because some of the stored water will evaporate. Evaporation rates in the Ethiopian highlands average about 1 meter a year. All together, the dam and irrigation projects envisioned in the Bureau of Reclamation plan would decrease the Nile's volume by 4–8 bcm a year.[11]

Moreover, Zewdie Abate, one of Ethiopia's top water and environmental officials, has suggested that his nation might use Nile water to develop substantially more irrigated land. "The implication," he writes, "that Ethiopia might utilize water to irrigate even half of the 3.7 million irrigable hectares is very significant to downstream users." Already, the Ethiopian government has hired international engineering firms to begin work on hundreds of small dams, mainly for irrigation. Some 200 of them, collectively able to store 477 mcm of Nile water a year, were reportedly completed by August 1997.[12]

Many small projects can have as great an impact on a river's flow as one large project, and there is good reason to believe these impacts could materialize. Ethiopia has a long tradition in small-scale agriculture, and many farmers have the experience and skills to build small dams in the highlands. Most donors would shy away from financing a large dam in Ethiopia because of negative impacts on Egypt and Sudan downstream, but microdams do not necessarily require outside technical expertise or funding. They also have strategic advantages. A single large dam presents Egypt with an obvious military target, whereas hundreds or thousands of small dams do not.

To take Abate's statement further, if Ethiopia decided to irrigate 1.8 million hectares, and if each hectare of crops consumed 4,000 cubic meters of Nile water annually, flows downstream could drop by 7.2 bcm a year, not counting evaporation from the thousands of small reservoirs created by the microdams. Even if a strategy of microdam construction could realistically irrigate only half as much area, a flow reduction of 3.6 bcm a year would still have a serious impact on Egypt.[13]

Meanwhile, Egypt continues to bring thirsty new irrigation projects on line that will make it more difficult to yield any water to Ethiopia. As noted in Chapter 4, in October 1997 Egyptian President Hosni Mubarak opened a major new canal to carry Nile water beneath the Suez Canal into the Sinai Desert. Begun under the administration of President Anwar Sadat, the canal is slated to irrigate 250,000 hectares of land. Earlier that year, Mubarak also gave the go-ahead for an ambitious scheme known as the Toshka or New Valley project, which would pump up to 5 bcm a year from the Nile upstream of Aswan Dam and send it by canal to Egypt's western desert, where it would irrigate some 200,000 hectares. With an estimated price tag of $87–145 billion over the next 20 years, the Toshka scheme is intended not only to expand irrigation but also to open new desert frontiers to human settlement, alleviating population pressures in the crowded Nile valley.[14]

Ethiopian officials have openly criticized the Toshka project. London's *Financial Times* quoted Mohamed Hagos, chief engineer in Ethiopia's Ministry of Water Resources, as saying that there can be no cooperation among the Nile countries "unless there is a clear commitment... to the principle of fair and equitable utilisation." Egyptian officials contend that the water for the Toshka project will come from stepped-up conservation, recycling, and reuse and will not cause Egypt to exceed its Nile allocation. But the project's water requirements will virtually eliminate any maneuvering room Egypt still has, making it much harder to strike a deal that accommodates Ethiopia's water needs.[15]

The historical context and current politics of the Nile basin make it hard to be optimistic about a peaceful resolution of looming water disputes. First, the region has a long history of hegemonic control over water. During the colonial period, the British dominated most of the Nile basin, including Egypt, Sudan, and the upper White Nile countries of Kenya, Uganda, and Tanzania. Ethiopia was colonized by Italy, and the Congo Free State, by Belgium. The boundary agreements established by the colonial powers around the turn of the century prohibited any water developments on the upper tributaries that might obstruct the Nile's flow into Sudan and Egypt without prior consultation and agreement.

Control passed to Egypt after it won independence from Britain in 1922. In 1929, the Egyptian government reached an agreement with the administrations of Sudan and East Africa (on behalf of the British government, which still controlled them) that recognized the need to develop irrigation in Sudan, but stipulated that any such development not infringe upon Egypt's historic rights to the river. Egypt assumed rights over the entire natural dry-season flow of the Nile, relegating Sudan's use to water stored at the end of the flood season. This 1929 agreement made no reference at all to the water rights of the East African countries.[16]

After World War II, as the basin states gradually won independence from their colonizers, the politics of the region shifted. A new administration in Sudan laid out plans to expand irrigation that required a greater share of Nile water. Meanwhile, Egypt was moving forward with plans for the Aswan High Dam, which would create a reservoir extending 150 kilometers into Sudan's territory. Tensions between the two nations mounted between 1954 and 1958, and after negotiations came to an impasse, Sudan declared it would no longer abide by the 1929 treaty.

After the military takeover in Sudan in late 1958, negotiations resumed and resulted in another bilateral treaty, signed in 1959. This established new Nile allocations for Egypt and Sudan based on the additional water to be made available by the massive reservoir behind the High Dam. As noted earlier, of the 84 billion cubic meters of Nile water that would enter the reservoir on average each year, 55.5 bcm would go to Egypt and 18.5 bcm to Sudan. The remaining 10 billion was expected to evaporate from the reservoir's surface. The new agreement diffused tensions between Egypt and Sudan, and allowed both nations to pursue their irrigation plans with an assured water supply. Because the treaty did not include any of the other basin states, however, and gave the two most downstream nations rights to the vast majority of the Nile's flow, upstream nations have never agreed to respect it.

At the moment, two divergent tracks are evident in the area. On the one hand, representatives from the basin countries have been meeting annually since 1993 to foster cooperation, and these meetings have borne some fruit. In addition, a group of water experts from 6 of the 10 Nile countries has been meeting regularly under the auspices of the U.N. Development Programme to work toward a legal and institutional framework for cooperation. In 1997, the 10 basin states appeared ready to embrace a cooperative approach, in part to secure $100 million in funding for joint water projects that the World Bank and prospective donors had made conditional on basinwide coop-

eration. Such trust-building activities and third-party mediation efforts can move otherwise hostile nations in the direction
of cooperative solutions.[17]

On the other hand, the water development projects that
Egypt and Ethiopia are pursuing unilaterally put the two
nations on a collision course. Egypt is increasing its dependence on Nile water just as Ethiopia is taking actions that will
reduce downstream flows. At a March 1998 meeting of the Nile
basin states in Tanzania, Ethiopia demanded a reexamination
of the 1959 treaty between Egypt and Sudan. Prior to that
meeting, the Ethiopian foreign minister reportedly said "the
time has come to erect dams and build reservoirs at the source
of the Nile in Ethiopia."[18]

Meanwhile, Sudan, if it ever emerges from the grip of civil
war, has plans to expand irrigation to another 1.5 million
hectares, and would like to build another dam on the Nile
north of Khartoum. All together, the Nile countries have plans
to bring irrigation water to an additional 2.9 million hectares.
Even if irrigators were highly efficient, this expansion could
easily require 25–35 billion cubic meters of additional water.
There is simply not enough water available in the Nile basin to
come close to achieving these irrigation goals.[19]

Moreover, some projects planned decades ago would likely
not pass international environmental scrutiny today. For
instance, Egypt and Sudan both continue to hold out hope for
a decades-old scheme to dredge and drain part of the Sudd
wetlands on the White Nile in southern Sudan. One of the
largest swamps in the world, the Sudd harbors a treasure trove
of wildlife—buffalo, elephant, gazelle, hippopotamus, zebra,
and several varieties of antelope. The Sudd wetlands are also
home to millions of migratory birds during the course of the
year, with the largest single population—the glossy ibis—numbering some 1.7 million during the dry season. A number of
Nilotic tribes—including the Dinkas, Nuer, and Shilluk—also
thrive in the Sudd. With populations collectively numbering

200,000–400,000, they key their lives of pastoralism, fishing, and agriculture to the seasonal cycle of flooding and the rich biological diversity of their homeland.[20]

The very vastness that makes the Sudd such an important and unique habitat, however, also makes it a source of huge evaporation losses. As the White Nile slows and spreads out across the swamp, some 34 bcm a year evaporate—a volume equal to 40 percent of the Nile's average annual flow at Aswan. Egypt and Sudan jointly planned a series of projects to channel water through the marshes so as to reduce evaporation and augment flows downstream. In accordance with their 1959 treaty, they agreed to share equally the additional water supplies the projects would generate.[21]

The initial phase of an early project called the Jonglei Canal, which was designed to capture an additional 4 bcm a year, was 70 percent complete when the Sudanese civil war broke out in 1983, halting work. Resurrecting the project at this point would be extremely costly. Moreover, international interest in protecting unique and important ecosystems like the Sudd has risen markedly during the last 15 years. As Tony Allan of the University of London and a specialist on Nile water politics points out, "no international or bilateral agency will finance the perceived impairment of one of Africa's major wetlands."[22]

Although Egypt has clearly dominated Nile water politics to date, the nation is probably not in a position to impose a water-sharing formula on the other basin states, especially if Ethiopia and Sudan should side together. As five members of the Middle East Water Commission put it in their book, *Core and Periphery*, "Egypt's nightmare is that Sudan and Ethiopia ally in the use of the Blue Nile." As recently as 1995, the government of Sudan threatened to cut the flow of Nile water into Egypt as tensions flared during a series of charges and countercharges surrounding the attempted assassination of Egyptian President Mubarak in Addis Ababa.[23]

Of all the countries in the basin, Egypt has the most to gain from a basinwide agreement on sharing the waters of the Nile. It is just a matter of time before Ethiopia and Sudan start to use more Nile water. Egypt will lose out if its upstream neighbors pursue their water schemes unilaterally. By undertaking its Toshka–New Valley scheme, however, Egypt sets the stage for precisely that scenario. That nation could take a productive step forward by abandoning Toshka and instead using the 5 billion cubic meters that were to go to that project to reach an agreement with Ethiopia.

Rules of Law

International law is only modestly helpful in guiding nations on the development and use of shared rivers. During this century, nations have signed 145 treaties dealing with non-navigational uses of water—an impressive number—but only one fifth of these treaties has any enforcement mechanism, and fewer than half have any monitoring provisions. Moreover, the vast majority of these treaties are between two countries, even in river basins shared by three or more nations.[24]

In May 1997, the United Nations General Assembly approved the long-awaited Convention on Non-Navigational Uses of International Watercourses. It is based on draft articles developed over 20 years by the U.N. International Law Commission. Although the Convention is a positive step, it is too vague and general to be of much practical help in hammering out water-sharing agreements.[25]

The Convention establishes two key principles to guide the conduct of nations over shared rivers—"equitable and reasonable use" and the obligation not to cause "significant harm" to neighbors. But these notions are open to widely differing interpretations. The Convention lists a number of criteria for nations to consider in deciding what is equitable and reasonable—including climate, geography, hydrology, population, existing and potential uses of the river, and the availability of alternatives—but offers no formula for weighting them. As a result, a nation in Egypt's position would place great weight on rights acquired through a long history of using the river, while one in Ethiopia's position would emphasize national contributions to the river's flow and future irrigation potential.[26]

Nonetheless, international acceptance of "equitable and reasonable use" as a guiding principle in shared river basins makes it more difficult for dominant upstream countries to completely ignore the needs of downstream countries or to inflict obviously inequitable arrangements on other basin states. Only three countries voted against the Convention—Burundi, China, and Turkey—and the latter two are powerful upstream countries with little fear of retaliation if their water development schemes harm others. Indeed, Turkey is one of the few countries that has occasionally still championed a now maligned principle called the Harmon Doctrine.[27]

Named after U.S. Attorney General Judson Harmon for an 1895 opinion he rendered in a U.S.-Mexico dispute over the Rio Grande, this doctrine essentially says that nations have absolute sovereignty over water within their borders and no obligation to share it with downstream neighbors. It has never really become part of international customary law, but occasionally an upstream state will invoke it. For example, in response to Syrian requests for more Euphrates River water, former Turkish Prime Minister Suleyman Demirel reportedly remarked in 1992: "We do not say we should share their oil resources. They cannot say they should share our water resources." On May 21, 1997, the day the water-sharing Convention was adopted at the U.N. General Assembly, the representative of Turkey explained that his country would not sign it and that it would have no legal effect in Turkey in part because it "did not refer to the sovereignty of the watercourse States over the parts of international watercourses located in their territory."[28]

Turkey and the Twin Rivers

Turkey's position in the Tigris-Euphrates basin is critical to the political stability and irrigation potential of the entire Middle East. Together, the Tigris and Euphrates rivers carry 50–60 percent as much water in an average year as the Nile does. Both rivers rise in the mountains of eastern Turkey. They then flow through Syria and Iraq, and join for a short distance before emptying into the Persian Gulf. (See Figure 7–2.)[29]

In its poor, less developed Southeast Anatolia region, Turkey is building a huge irrigation and hydropower scheme known as the GAP (from the name in Turkish). The GAP includes the construction of two dozen dams, and aims to irrigate 1.7 million hectares of land, generate 27 billion kilowatt-hours of electricity, boost the region's income fivefold, and generate jobs for 3.5 million people. Syria and Iraq fear that the $32-billion scheme will foil their own development plans. If

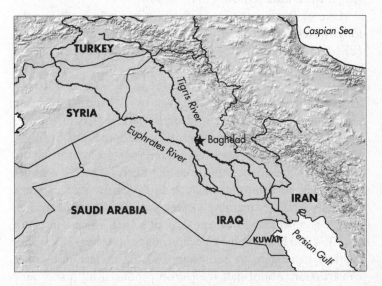

Figure 7–2. Tigris and Euphrates River Basins

completed as planned, the GAP could reduce the Euphrates flow into Syria by 35 percent in normal years and substantially more in dry ones, besides polluting the river with irrigation drainage. Iraq, third in line for Euphrates water, would see a drop as well.[30]

Syria and Iraq have agreed to share whatever mainstream Euphrates water crosses the Turkish-Syrian border, with Syria getting 42 percent and Iraq getting 58 percent. But both nations want assurances of more water. Turkey and Syria signed a protocol in 1987 that guarantees the latter nation a minimum flow of 500 cubic meters per second, about half of the Euphrates's volume at the border. But Syria wants more, and has also complained about the reliability and quality of water it is receiving. Some Israelis would see resolution of the Euphrates dispute as beneficial to Israel's position in the Jordan basin, as this might relieve some pressure in future negotiations over the Jordan headwaters and the Golan Heights, which Israel annexed from Syria in 1981.[31]

One of the few trump cards Syria has held is its decade-long, widely reported support of the Kurdish Workers Party (PKK) in that group's 14-year separatist war in southeastern Turkey—support that Turkey has long wanted stopped. Tensions between Turkey and Syria flared in October 1998, mainly over the Kurd issue, but also over Turkey's planned diversions of the Euphrates. Turkey warned that differences with Syria had reached the stage of "undeclared war."[32]

In late 1998, Syria reportedly agreed to stop supporting the PKK. The subsequent capture of the Kurdish guerrilla leader who had lived somewhat secretly in Syria for years has now reduced Syria's bargaining power with Turkey. Meanwhile, Turkey is proceeding with the GAP at a fairly rapid clip. More than 174,000 hectares have been brought under irrigation, and facilities to irrigate an additional 183,700 hectares are under construction. About half the cotton produced in Turkey now comes from the GAP region. Financing, however, remains a problem. The World Bank will not finance the scheme in the absence of a water-sharing agreement with Syria and Iraq. Since other lenders often follow the Bank's lead on these matters, Turkey has had limited success raising international funds. The Turkish government has spent $12.6 billion—some 38 percent of the project's estimated total cost—but the balance has yet to be raised.[33]

Toward Win-Win Water-Sharing

Breaking the logjams in these three Middle East river basins is critical to regional stability. It is also key to maximizing the total benefits derived from each river. Politics has locked these nations into a zero-sum game, in which one party's gain is perceived as another's loss. Through cooperation, however, there are a variety of possible win-win strategies, in which all parties benefit. Historical experience suggests several useful ideas for moving in this direction.

First, shifting the debate from discussions of water rights to water needs can prove constructive. How much water each party has a "right" to is subjective and culturally and emotionally charged, and it varies greatly with the criteria used. How much water each party "needs" or can beneficially use, however, can be quantified more objectively. Such a needs-based water-sharing formula was developed for the Jordan basin in the 1950s, and although now somewhat anachronistic, this set of terms brought the region closer to a basinwide water agreement than it has ever been since.

In 1953, five years after Israel attained nationhood, U.S. President Dwight D. Eisenhower appointed Eric Johnston as special ambassador to the Jordan basin to help negotiate the region's water development. The so-called Johnston formula involved estimating how much water was needed for all of the potentially irrigable land within the basin that could receive surface water by gravity. The architects of the plan then put existing political boundaries on top of this map of irrigation potential, and in this way determined how much water each nation would receive. Although irrigable area was the basis for determining the allocations, the plan allowed each country to use its share of water any way it pleased.

Because the Johnston plan was technically feasible, met the specified needs of each nation in an equitable way, and yielded obvious benefits for the whole basin, it was acceptable to all parties at the time. (It did not deal with groundwater, however, which has become a vital and highly disputed source of supply for both Israel and the West Bank. Nor did it include the Palestinians as a distinct political entity.) Politics presided over rationality, however, and the parties never formally ratified the plan. But the Johnston formula served as a guide for the basin's future water development and provided hope that the parties might at some point reach a binding agreement.[34]

In addition to showing the value of focusing on water needs rather than rights, the Johnston plan underscored the importance of reaching agreements before a region's water predicament becomes too tight. During the four decades since the formula was devised, water conditions in the Jordan basin have changed markedly as a result of population growth, higher water demands, and wars that altered both the regional balance of power and control over water. Israel, for example, has been taking about 50 percent more surface water from the basin than the Johnston formula specified, while Jordan, which is desperately water-short, currently gets less than its Johnston allotment.[35]

A third important lesson that emerges from past experience is that mediation by an outside party is often key to resolving water disputes, and that this mediation may need to be backed by financial support. After brokering the Johnston plan, the United States provided aid to Israel and Jordan as they proceeded on their separate paths of water development, which initially were based on the Johnson formula. Similarly, the

World Bank played a key intermediary role in resolving a 12-year dispute between India and Pakistan over the Indus River, a disagreement that erupted when the subcontinent was partitioned in 1947. World Bank President Eugene Black used his good offices to help steer the two nations away from armed conflict and toward a water-sharing agreement, which was reached in 1960. The Bank also mobilized financing to build the water infrastructure that was integral to the success of the Indus Waters Treaty. This degree of leadership by an outside party has not occurred since.[36]

Fourth, a fortuitous shift in political climate can open up possibilities for breaking a long-standing logjam, provided negotiators are prepared to take advantage of it. For example, Bangladesh and India spent years deadlocked in a dispute over the dry-season flow of the Ganges River, which originates in Nepal and flows more than 2,200 kilometers through the other two nations before emptying into the Bay of Bengal. (See Figure 7–3). Of all the disputed rivers, none affects so many people as the Ganges—several hundred million in all. In the early 1970s, India built a barrage at Farakka, not far from the Bangladesh border, in order to divert Ganges water to Calcutta and thereby improve navigation conditions in this port city. Newly independent Bangladesh became concerned that not enough river water would cross into its territory during the dry season, and that its crop production would suffer.

The two nations signed a series of short-term agreements, but the last of these expired in 1988, leaving Bangladesh in a precarious position. Tensions flared in 1993, when the river dropped to its lowest recorded level, idling Bangladeshi irrigation pumps and causing severe crop losses. Two years later, Bangladeshi Prime Minister Begum Khaleda Zia went before the United Nations and called India's heavy diversions near the border "a gross violation of human rights and justice," and hailed the Farakka Barrage "an issue of life and death" for Bangladeshis.[37]

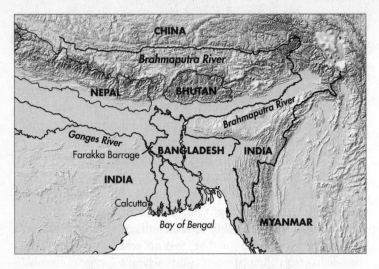

Figure 7–3. Lower Ganges River Basin

Then in December 1996 the dispute dissipated with the signing of a water-sharing treaty by the two governments. The agreement came as a surprise to many water observers, because India—as both the upstream nation and the stronger one— seemed to have little to gain from bargaining. But new possibilities for conciliation opened up with political changes in both countries: in India, with the "good-neighbor policy" of a coalition government headed by H.D. Deve Gowda and the diplomatic efforts of his external-affairs minister, Inder Gujral (who shortly thereafter became India's prime minister); in Bangladesh, with the return to power after 21 years of the pro-India party led by Prime Minister Sheikh Hasina.[38]

The new treaty, which is to remain in force for 30 years, grants Bangladesh about three times more Ganges water than it had been receiving. (See Table 7–4.) It establishes a joint commission to monitor flows at Farakka and to carry out the treaty's provisions. It also includes a guaranteed minimum flow to Bangladesh in the event that disagreements arise. India got nothing specific in return for its good will, but hopes that Bangladesh might extend the spirit of cooperation by granting concessions on transit rights (Bangladesh has not allowed goods and people to cross its territory between India's north-eastern states and West Bengal), immigration, or other issues. Indeed, the massive flow of illegal immigrants into India from Bangladesh's worst-hit district, Khulna, no doubt explains some of India's motivation to negotiate.[39]

For its part, Bangladesh now has sufficient water security to proceed with plans to get more benefit out of the river. The government is examining the feasibility of building its own barrage at Farakka to augment dry-season water supplies and to slow the advancing salinization that is damaging farmland, fisheries, and mangrove forests in the Ganges delta. Estimated to cost about $1.5 billion, the barrage is widely viewed as a keystone to Bangladesh's development and to the protection of the lower Ganges ecosystem. At an international gathering of water experts and financial donors in March 1998, Prime Minister Hasina underscored the importance of the new treaty to her nation's ability to optimize use of the Ganges and improve the

Table 7–4. Terms of 1996 Water-Sharing Agreement Between Bangladesh and India

Ganges Flow at Farakka[1]	Share to India	Share to Bangladesh
70,000 Cusecs or Less	50%	50%
70,000-75,000 Cusecs	Remainder of Flow	35,000 Cusecs
75,000 Cusecs or More	40,000 Cusecs	Remainder of Flow

[1]A cusec is a measure of flow equal to a cubic foot (or 28.3 liters) of water per second.
SOURCE: Treaty Between the Government of the Republic of India and the Government of the People's Republic of Bangladesh on Sharing of the Ganga/Ganges Waters at Farakka, Annex 1, 12 December 1996.

livelihoods of the 35 million Bangladeshis who depend on the river: "We could not do this in the past," she said, "due to lack of a long term water sharing agreement with India."[40]

Ultimately, the only way for nations in water-scarce river basins to maximize the benefits derived from their shared river is to cooperate in the joint management of the river basin as a whole. To date, examples of this degree of cooperation are rare. But as deepening water scarcity crimps economic development, causes food import bills to rise, and exacts higher ecological costs, the compelling logic of cooperation may begin to override the stubborn politics of zero-sum posturing. In many cases, the seeds of cooperation lie precisely within the very sources of conflict that make win-win solutions appear so elusive.

At the heart of many river disputes is a conflict over acquired rights to water that derive from historical use of a river versus the future economic potential of that river in countries that have used comparatively little of its water to date. In the Nile basin, as noted earlier, Egypt has drawn upon the river for 5,000 years, steadily claiming more water as population and food demands increased. More recently, Ethiopia and Sudan have expressed desires for additional Nile water to advance their own economies and food-production capacities.

From a regional perspective, storing water in Ethiopia makes much more sense than storage in Egypt. Fully 12 percent of the Nile's volume—an average of 10 billion cubic meters a year—evaporates from Lake Nasser under the hot desert sun. Storing water in the Ethiopian highlands, where the evaporation rate is one third as high as at Aswan, would increase the volume of water available for use within the basin, potentially benefiting all parties. Although any major dam construction in the highlands would cause some environmental damage, this could be lessened by operating the dams in ways that help restore and protect ecological values and functions downstream. (See Chapter 6.) By some estimates, the possible water savings from shifting storage sites equals the total volume of water required by all the irrigation projects included in the Bureau of Reclamation's plan for the upper Blue Nile. Viewed another way, if shifting storage to upstream reservoirs cut annual evaporation losses at Aswan in half, the saved water would be sufficient to grow an additional 4 million tons of grain—about half of Egypt's annual net grain imports.[41]

Another opportunity for win-win savings exists on the White Nile. Sudan's Jebel Aulia Dam, about 40 kilometers south of Khartoum, loses half of the water it is capable of storing to evaporation, about 2.8 billion cubic meters a year. Since construction of the High Dam at Aswan, the storage provided by Jebel Aulia Dam is no longer fully needed. The dam's main purpose now is to raise the water level along the White Nile in order to reduce the pumping lift of downstream irrigation schemes, covering an area of about 40,000 hectares. Rehabilitating the pumps so that they can draw water year-round regardless of the river's level would eliminate the need for so much storage behind Jebel Aulia Dam. This in turn would greatly cut evaporation losses, freeing up perhaps 1 bcm or more a year for irrigation and other uses downstream. In the absence of basinwide cooperation and because of its enduring civil war, Sudan has not obtained financial assistance for this rehabilitation effort.[42]

Cooperation among all the Nile basin countries—but especially Egypt, Ethiopia, and Sudan—offers the most promising set of win-win options. The challenge is to arrive at a pact that creates mutual interdependence based on each nation's comparative advantage—an arrangement that both binds the basin states together and optimizes the use of land, water, technical knowledge, and other resources. A variety of win-win arrangements can be generated by capitalizing on Ethiopia's undeveloped headwaters, Sudan's fertile and irrigable soils, and Egypt's technical skill, industrial development, and ability to mobilize financing.

For example, water-sharing agreements could include regional food-security arrangements. Egypt might find it beneficial to allow some of its Nile water to be used for crop cultivation in Sudan in exchange for some of the food produced. Joint investments to reduce evaporation losses—for instance, by altering water storage locations or investing in irrigation efficiency improvements—could save water to the benefit of all parties, possibly eliminating the need for projects that would harm the Sudd. These kinds of arrangements would allow irrigators in the basin to grow more food with Nile water, lessen the region's dependence on world grain markets, and protect some of their environmental assets.

A 1925 treaty dividing the waters of the Gash River, which flows from Eritrea into Sudan, incorporated this kind of cooperative approach. Signed by the two colonial powers, Italy (for Eritrea) and the United Kingdom (for Sudan), the pact included a provision that Sudan pay Eritrea a share of the income derived from cultivation in the Gash delta—specifically, 20 percent of any sales over 50,000 pounds. While the treaty overall was skewed in Eritrea's favor, the principle of sharing in the benefits derived from the river rather than sharing the river itself is an important one. It opens up numerous possibilities for creativity in maximizing total basinwide benefits from a river's management and use.[43]

Each layer of cooperation among countries requires deeper levels of trust, and in hot spots of water dispute, this trust is difficult to achieve. Even if cooperation is obviously beneficial to all parties over the long term, political leaders have a short-term incentive not to join in because of the risk that other parties will not live up to their end of the bargain. Nevertheless, in water-scarce regions such as the Nile, the logic of some form of basinwide agreement is fast becoming inescapable. Strategically, it is in Egypt's interest to negotiate a pact before Sudan and Ethiopia team up to develop the Blue Nile jointly. It is in Ethiopia's interest to reach an agreement, because it will need international financing to implement projects in the upper Nile watershed. And it is in Sudan's interest because the nation has substantial irrigation potential that requires additional water storage, which may best be done in Ethiopia.[44]

Because the gains from cooperation are often highly asymmetrical, negotiators may need to consider goods and resources other than water. For example, the Nile plays a relatively small part in the water supply picture of the upper White Nile countries, so these nations may only support a basinwide water-sharing agreement if they secure some other benefits in return. Likewise, an agreement signed by three of the Aral Sea basin countries in April 1996 calls for upstream Kyrgyzstan to store more of the Syr Darya's flow in the winter for release in the spring, when downstream Uzbekistan and Kazakhstan need irrigation water for the planting season. Because this will cut Kyrgyzstan's winter hydroelectric output, the deal calls for the nation to receive in return some of Uzbekistan's natural gas and Kazakhstan's coal. Brokered by the U.S. Agency for International Development, the agreement—if implemented—resolves only a small part of Central Asia's water problems, but it is a promising step toward cooperation.[45]

Finally, the inevitability of droughts and the prospect of altered river runoff as a consequence of global warming call into question the soundness of treaties that allocate fixed quantities of water to each party. In many years, there may simply not be enough water in the river to satisfy all treaty commitments. A more sensible approach is for agreements to specify each party's respective share of river runoff, allowing the actual quantities of water to vary with the flow that year.[46]

To protect a river's ecological functions, treaties would also need to specify an absolute quantity and quality of water that is reserved for the river system itself, including its delta. This environmental allocation would always need to be satisfied first. The parties to the treaty would then share the remaining water. The situation in the Colorado River basin underscores why this kind of approach is important. The two key treaties that divide the Colorado among seven states and Mexico allocate more water than the river actually carries in an average year—a mistake that occurred because the river's annual flow was determined in an unusually rainy period. As a result, virtually no fresh water flows through the Colorado delta and into the Sea of Cortez in an average year. The Colorado delta in northern Mexico, and the native Indian communities that live there, have been decimated. Had the treaty designers set aside fresh water to maintain this critical downstream ecosystem, they might have prevented substantial ecological and social harm.[47]

It may be pointless to predict whether water wars will headline the news in the twenty-first century. But it is far from pointless to try to prevent those wars from occurring. Whether international or domestic, tensions over water scarcity have the potential to incite civil unrest, spur migration, impoverish already poor regions, and destabilize governments—as well as to ignite armed conflict.

In recent years, the intelligence community has awakened to these risks and has broadened its definition of security to include environmental deterioration and resource scarcity. In a landmark speech at Stanford University in April 1996, U.S. Secretary of State Warren Christopher proclaimed that the United States must contend with "the vast new danger posed to our national interests by damage to the environment and resulting global and regional instability." He cited the need for regional strategies to confront pollution and resource scarcity where they increase tensions within and among nations. Nowhere is this more evident," he said, "than in the parched valleys of the Middle East, where the struggle for water has a direct impact on security and stability."[48]

A little over a year later, the State Department set up six regional "environmental hubs" around the world, and four of them have water as a top priority: Amman, Jordan, in the Jordan River basin; Tashkent, Uzbekistan, in the Aral Sea basin; Kathmandu, Nepal, in the headwaters of the Ganges River; and Addis Ababa, Ethiopia, in the upper Nile watershed. These developments come none too soon. Water-sharing agreements are difficult to hammer out, and often take many years of negotiations. But as U.S. Deputy Secretary of State Strobe Talbott remarked in 1996, "We are beginning to understand, perhaps for the first time, the sometimes devastating, sometimes promising, always complicating interaction between human history and natural history."[49]

8

THE PRODUCTIVITY
FRONTIER

*One of the most fateful errors of our age is the belief that the
problem of production has been solved.*

E. F. Schumacher

In many ways, irrigated agriculture is stuck between a rock and
a hard place. Because of limited opportunities to get more food
from rain-fed croplands, we are counting on irrigated lands to
provide the bulk of the additional food needed in the decades
ahead. But at the same time, because of dwindling options to
expand water supplies, we are hoping to shift water away from
agriculture to satisfy rapidly growing urban and industrial
demands, restore fish populations, safeguard endangered
species, and protect the critical ecological functions of rivers
and wetlands. Can we reconcile these competing interests? Or
will one or more of them suffer?

In large part, the answer will hinge on our ability to do
more with less water. For most of the modern irrigation age,
water has seemed plentiful and cheap, so farmers, researchers,
and engineers did not put much effort into growing food in
water-thrifty ways. Now that water is increasingly scarce, however, raising water productivity—getting more service, satisfac-

tion, and benefit out of every liter we remove from a river, lake, mountain spring, or underground aquifer—is a key to meeting our future needs. Just as land productivity—the amount of crops produced per hectare of land—became the frontier to exploit during the latter half of the twentieth century, so water productivity—getting more crop per drop—is the agricultural frontier for the twenty-first century.

Orchestrating this "Blue Revolution" will be more difficult than the Green Revolution of the past several decades. Along with new high-yielding crop varieties, the Green Revolution depended on increased use of agricultural chemicals and irrigation water for its success. Without fertilizer, the new seeds could not reach their yield potential, and without sufficient water, the fertilizer would have little effect. The 2.4-fold rise in world grainland productivity between 1950 and 1995 was matched by a 2.2-fold rise in irrigation water use. (See Figure 8–1.) So although the Green Revolution helped conserve land by sparing substantial areas of forest and grassland from the plow, it demanded vast quantities of water.[1]

In contrast, there is no obvious, off-the-shelf package available to raise water productivity. This new challenge will require a more diverse and creative mix of strategies that together make agriculture more information-intensive and less resource-intensive—in most cases, by substituting technology and better management for water. It will require thinking systemically, because water performs many different functions as it flows through the landscape toward the sea. Diverting or consuming water in one place has an impact someplace else. Even gauging progress toward greater water productivity is fraught with problems since we rarely measure water use accurately. For example, no international agency systematically records what crops farmers are growing on irrigated lands, much less how much water they apply there.

Nevertheless, farmers, researchers, engineers, and others around the world are beginning to paint the outlines of more

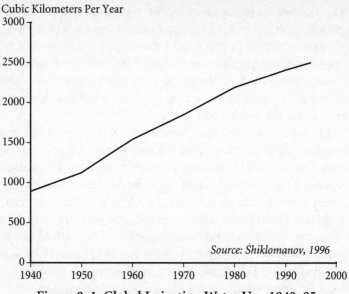

Figure 8–1. Global Irrigation Water Use, 1940–95

water-conserving methods of meeting global food needs. There is no magic bullet, no quick fix to agriculture's water problems. But the technologies and strategies described in this chapter inspire hope that we can achieve the doubling of water productivity needed to satisfy the food, water, and environmental needs of the next several decades—if we so choose.

Tracking Water

A person plopped down into a random field of irrigated grain would likely have a hard time knowing what century it is. Most farmers today irrigate the way their predecessors did hundreds of years ago. Water is delivered to the farm by canal. Farmers then flood their fields or channel the water down parallel furrows, letting gravity move the water from one end of the field to the other. Along the way, some of the water ends up in the root zone of the soil, some of it percolates past the root zone more deeply into the subsurface, some of it runs off the tail-end of the field, and some of it evaporates. Only the water that replenishes the root zone (plus some extra to flush salts away) actually helps the farmer's crops grow. The remainder—often half or more of the water applied—leaves that field without having produced any benefits.

In one sense, this "inefficiency" could be good news, because it suggests a large potential for saving water. In reality, however, much of this "wasted" water gets used by another farmer, factory, city, or ecosystem further downstream. Consider, for example, the basin irrigation that sustained ancient Egypt for thousands of years. (See Chapter 2.) The system was not very efficient from the perspective of each individual farm plot, because much of the water spread over each field drained off without benefiting a crop. However, since the runoff from upstream farms became the supply for downstream farms, the system's efficiency was much higher from the perspective of the whole Nile valley.

Just about any improvement in irrigation efficiency helps protect the environment, because the less water farmers apply to croplands, the fewer salts and pollutants get added to local water bodies. Actually saving water, however, is more complicated.

To help make this point, David Seckler, Director General of the International Water Management Institute in Sri Lanka, draws a distinction between "wet" (or real) water savings and "dry" (or paper) savings. "Wet" savings actually generate new water supplies. For example, reducing the amount of water that evaporates from the soil or from irrigation canals keeps more water in the river basin, and thereby increases the available supply. "Dry" savings, on the other hand, merely shift water around. For instance, most of the water that the Imperial Irrigation District in California plans to "save" and sell to San Diego (see Chapter 6) now flows to the Salton Sea, where it helps sustain the lake's level and volume. It may make sense to line the irrigation district's canals and capture this water if it contributes more value to society in San Diego than in the Salton Sea. But this kind of action does not save water or generate new supplies from the perspective of the whole river basin. It merely shifts water from one user to another—a case of robbing Peter to pay Paul.[2]

Saving water, then, is not as simple as it seems. It requires knowing where water goes and what functions it performs. It also depends on the geographic scale or vantage point we are concerned with. From a planetary perspective, it is impossible to save water because none is ever wasted. Earth is a big water-recycling machine, moving water between the land, the sea, and the atmosphere. No water gets lost, it merely changes location, quality, and form. To a river basin manager, however, water certainly can be wasted. If water evaporates from an expensive reservoir or from a farmer's field, it leaves human control without having produced any benefits. Finally, an individual farmer who pays to pump water from a well might view any water that flows off the farm as a waste, even though that water may supply someone else downstream.

From society's standpoint, there are four major ways to improve water productivity. First, we can reduce the portion of our developed water supply that is lost to evaporation and unrecycled seepage, and then put those water savings to productive use—whether that be growing another crop, supplying a factory, or supporting a wetland. Second, we can develop and spread new crops, cultivation methods, and irrigation practices that allow farmers to get a bigger harvest from the same volume of water.

Third, we can capture more fresh water before it flows into "salt sinks"—oceans, inland lakes and seas, or other salty water bodies. The most familiar example of this strategy is the building of dams to capture and store river water that would otherwise run rapidly to the sea in floods. And fourth, we can shift water from one use to another that is considered more productive. Farmers can grow wheat instead of a thirsty nonfood crop like cotton, for example, getting more food and nutritional value out of that water. Or they can fallow some irrigated land and sell the saved water to a factory. This may increase the economic productivity of the water, and may have environmental benefits if it prevents the construction of another dam, but it reduces the volume of water used to grow food.

Each of these options for improving water productivity involves costs and benefits that vary from place to place. Dams and other ways of capturing more runoff are discussed in Chapter 4, and the shift of water out of agriculture to satisfy other demands is a topic of Chapter 6. This chapter focuses mainly on the first two options—getting more benefit out of irrigation water.

Counting Our Losses

No one knows how much irrigation water is lost to evaporation worldwide. In the hot, dry climates and seasons in which most irrigation takes place, water evaporates from the soil, from irrigation canals, from plant leaves and other wet surfaces—but in an amount we can only guess. When planners develop a water budget—an accounting of how much water flows into an area, how that water gets used, and then how it leaves the area—they often lump evaporation together with transpiration, the process by which plants take up moisture through their roots and then release it through their leaves to the atmosphere. They typically talk of water consumed through evapotranspiration, or ET.

They join the terms for convenience: the two processes are difficult to measure separately, and both return water to the atmosphere, rendering it unavailable for any additional use. Yet putting them together leads to a very misleading picture of how efficient our irrigation practices are. Transpiration is a necessary and beneficial use of water; it enables plants to grow. In contrast, most evaporation of irrigation water is not beneficial. By considering all water consumed through ET as productive, we overlook an important source of waste—and of potential water savings.

In developing water accounts for an irrigation system in India, David Molden of the International Water Management Institute made an effort to estimate transpiration and unproductive evaporation separately. He found that 16 percent of the supply available to irrigate wheat ended up being lost to evaporation. (Since rainfall accounted for just over a tenth of the wheat's total water supply, the actual share of irrigation water lost to evaporation was likely somewhat less.)[3]

Globally, if 10–20 percent of all the irrigation water pumped from wells or supplied from reservoirs ends up evaporating because of inefficient irrigation practices, an additional 250–500 billion cubic meters of water could conceivably be captured for use. Doing so would boost agriculture's water productivity substantially, and make additional water available to grow more crops or supply other activities. Five hundred billion cubic meters is roughly the amount needed to grow 500 million tons of grain—about a fourth of the current grain harvest. This volume could also meet 85 percent of the projected increase in household and industrial water demands in all developing countries by 2020.[4]

In addition to evaporation losses, some irrigation water seeps deeper into the soil and may take weeks, months, or even years to become accessible again, depending on the regional hydrology. Though not "lost" from the system, it is no longer available for use—at least for some time. A substantial quantity of irrigation water also becomes unfit for further use because it picks up too many pollutants or salts as it moves through the soil. Salty and toxic drainage is causing serious environmental problems in many irrigated areas, including California's Central Valley. (See Chapter 5.) By irrigating more efficiently, farmers can avoid these kinds of water losses and reduce damage to the environment—again, boosting water's productivity by making it available for more uses.

A long and growing list of measures can help farmers, water managers, and society as a whole get more output and benefit from water dedicated to irrigation. (See Table 8–1.) The key is to custom-design strategies to fit the farming culture, climate, hydrology, crop choices, water use patterns, environmental conditions, and other characteristics of each particular area. Successful strategies almost always involve a synergistic mix of measures from the water-productivity toolbox. Farmers will not invest in efficient technologies, for example, if they have no economic incentive to do so. And these technologies will only improve water productivity if accompanied by good management practices.

A Micro Revolution

Of the many ways available to raise agriculture's water productivity, drip irrigation ranks near the top of those with substantial untapped potential. Just as the creation of concrete opened the door for dams of unprecedented size and strength during the first half of the twentieth century, the development of inexpensive, weather-resistant plastic after World War II paved the way for another revolution in water use—this one at a micro scale.

Using a network of perforated plastic tubing installed on or below the soil surface, drip systems deliver water directly to the roots of individual plants. Much like a child's Lego building blocks, the network of tubing can be configured in various ways to match the pattern of crops in the field. The tubing can be left in place, or picked up and moved to different locations. Delivered under low pressure, the water emerges through small holes or emitters at a slow but steady rate, drop by drop. Drip irrigation's early designers adapted filters initially developed by the swimming pool industry to prevent particles or algae from clogging the dripper holes. When combined with soil moisture

Table 8–1. Menu of Options for Improving Irrigation Water Productivity

Category	Option or Measure
Technical	• Land leveling to apply water more uniformly • Surge irrigation to improve water distribution • Efficient sprinklers to apply water more uniformly • Low energy precision application sprinklers to cut evaporation and wind drift losses • Furrow diking to promote soil infiltration and reduce runoff • Drip irrigation to cut evaporation and other water losses and to increase crop yields
Managerial	• Better irrigation scheduling • Improving canal operations for timely deliveries • Applying water when most crucial to a crop's yield • Water-conserving tillage and field preparation methods • Better maintenance of canals and equipment • Recycling drainage and tail water
Institutional	• Establishing water user organizations for better involvement of farmers and collection of fees • Reducing irrigation subsidies and/or introducing conservation-oriented pricing • Establishing legal framework for efficient and equitable water markets • Fostering rural infrastructure for private-sector dissemination of efficient technologies • Better training and extension efforts
Agronomic	• Selecting crop varieties with high yields per liter of transpired water • Intercropping to maximize use of soil moisture • Better matching crops to climate conditions and the quality of water available • Sequencing crops to maximize output under conditions of soil and water salinity • Selecting drought-tolerant crops where water is scarce or unreliable • Breeding water-efficient crop varieties

SOURCES: Amy L. Vickers, *Handbook of Water Use and Conservation* (Boca Raton, FL: Lewis Publishers, in press); J.S. Wallace and C.H. Batchelor, "Managing Water Resources for Crop Production," *Philosophical Transactions of the Royal Society of London: Biological Sciences*, vol. 352, pp. 937–47 (1997).

monitoring or other ways of assessing crops' water needs accurately, drip irrigation can achieve efficiencies as high as 95 percent. Losses to evaporation are negligible, and because less water is pumped, less energy is used as well. Farmers can also apply fertilizer through the drip system, reducing chemical costs and the potential for land and water pollution.

Besides saving water, energy, and chemicals, drip systems usually offer a surprising bonus—higher yields. Before this system was developed, the guiding approach to irrigation was to apply as much water as the soil could hold, let the crops deplete that moisture almost to the point that they begin to wilt, and then irrigate again to refill the soil moisture reservoir. This traditional way of irrigating creates a kind of boom-and-bust cycle for the crop—a time of abundant moisture followed by a time of dryness—that is not typically good for the plant. Saturating the soil often starves crops of oxygen and leaches nutrients out of their root zone. Letting the soil dry out may damage the crop's roots.

Soil scientist Daniel Hillel, who participated in drip's development in Israel during the 1960s, likens this traditional irrigation philosophy to force-feeding babies a huge meal every Sunday and then waiting until they are practically famished a week later before feeding them again. Babies fed smaller amounts of food more regularly would clearly fare better. Similarly, drip irrigation is like "spoon-feeding" crops small amounts of nourishment regularly so that they are neither overfed nor underfed at any time. As a result, the plants are healthier, and farmers get higher yields and better-quality harvests.[5]

Drip and other so-called microirrigation techniques (mainly microsprinklers) were developed commercially in Israel. By the mid-1970s, farmers in a half-dozen countries—Australia, Israel, Mexico, New Zealand, South Africa, and the United States—were using drip methods on a portion of their cropland, and drip area worldwide totaled about 56,000 hectares. From this small base, microirrigation then spread rapidly, reaching nearly 1.6 million hectares by 1991. With the next survey of its extent not expected until after 2000, no one knows what the microirrigation area is today. But Dale Bucks, a water specialist with the U.S. Department of Agriculture who has overseen past surveys for the International Commission on Irrigation and Drainage, estimates that the global area has probably expanded by about 75 percent since 1991. If so, the area under microirrigation worldwide now totals about 2.8 million hectares.[6]

Although this growth is impressive—a 50-fold increase over the last two decades—drip and other highly efficient microirrigation methods still represent only 1 percent of the world's total irrigated area. In light of worsening water scarcity and rising water costs, several recent developments could lead to a major expansion in the area watered by drip irrigation.

First, researchers are collecting and disseminating the results of field experiments with drip irrigation more widely. In countries as diverse as India, Israel, Jordan, Spain, and the United States, studies have consistently shown drip irrigation to cut water use by 30–70 percent and to increase crop yields by 20–90 percent. The combination not only produces dramatic gains in water productivity—crop yield per unit of water—it markedly improves the economics of drip irrigation. Drip systems typically cost $1,200–2,500 per hectare, which most farmers judge to be too expensive, especially if their farm water is heavily subsidized. Factoring in the potential for higher crop yields, however, sharply lessens the payback period on a drip-system investment.[7]

Although Israel documented and began benefiting from drip's potential more than two decades ago, other countries have only more recently discovered the benefits for their own agricultural economies. India, for example, had only 1,000 hectares under drip irrigation as recently as 1985. Over the next decade, however, drip spread rapidly, reaching 71,000 hectares in 1994. R.K. Sivanappan, chair of the Centre of Agricultural Rural Development and Environmental Studies in Coimbatore and often called the father of drip irrigation in India, estimates that in just the last four years the area under drip has more than doubled, to 162,000 hectares. A tireless advocate of the technology, he believes drip's potential in India may exceed 10 million hectares—an area equal to 20 percent of the nation's current irrigated land.[8]

Studies by a number of Indian research institutions on a variety of crops have shown drip to cut water use by 30–60 percent and to boost yields by 5–50 percent over conventional surface irrigation. (See Table 8–2.) Unfortunately, these results do

Table 8–2. Water Productivity Gains from Shifting to Drip from Conventional Surface Irrigation in India[1]

Crop	Change in Yield	Change in Water Use	Change in Water Productivity[2]
	(percent)		
Banana	+52	− 45	+173
Cabbage	+ 2	− 60	+150
Cotton	+27	− 53	+169
Cotton	+25	− 60	+255
Grapes	+23	− 48	+134
Potato	+46	~ 0	+ 46
Sugarcane	+ 6	− 60	+163
Sugarcane	+20	− 30	+ 70
Sugarcane	+29	− 47	+ 91
Sugarcane	+33	− 65	+205
Sweet potato	+39	− 60	+243
Tomato	+ 5	− 27	+ 49
Tomato	+50	− 39	+145

[1]Results from various Indian research institutes. [2]Measured as crop yield per unit of water supplied.
SOURCE: Adapted from data in Indian National Committee on Irrigation and Drainage, *Drip Irrigation in India* (New Delhi: 1994); R. K. Sivanappan, "Prospects of Micro-Irrigation in India," *Irrigation and Drainage Systems*, vol. 8, pp. 49-58 (1994).

not distinguish how much of the savings are from reduced evaporation versus seepage or tail water that conceivably could be reused elsewhere. Nonetheless, as measured on the farm, the combination of water savings and yield increases typically results in more than a doubling of water productivity.[9]

A second reason farmers may turn more rapidly to drip irrigation is that researchers and irrigators are getting favorable results with two crops that are both thirsty and widely planted—cotton and sugarcane. In the United States, for example, Texas—which produces about 40 percent of the U.S. cotton crop—has gotten promising results with drip irrigation of cotton. The state's average cotton yields are low, in part because farmers lack sufficient water during July and August, when the crop needs a lot of water. In 1983, agents with the Texas Agricultural Extension Service paid a visit to Casa Grande, Arizona, to see the drip irrigation practices at Sundance Farms, which is owned by a creative innovator named Howard Wuertz.[10]

Wuertz pioneered a farming system that combines drip irrigation with minimum tillage of the soil. He buried drip tubing 8–10 inches deep in every crop row, and then practiced multiple cropping of vegetables and field crops (including cotton) along with minimum tillage, leaving the drip system in place. Studies of Wuertz's low-till drip methods by the University of Arizona showed that the system was able to cut water and energy use by about half and field labor by nearly 60 percent, while increasing lint yield from cotton crops by 13 percent.[11]

Impressed with what they saw, the Texas extension agents adapted the system for their local conditions back home. Several years of recordkeeping showed that drip-irrigated cotton was producing an average of 27 percent more lint per hectare than conventional furrow-irrigated farms in the same region. Although substantial water savings did not materialize in the Texas operations, the jump in yield greatly increased output per unit of water.[12]

More recently, the High Plains Underground Conservation District in Lubbock, Texas, a 15-county farm area that depends heavily on the dwindling Ogallala aquifer, has also begun experimenting with drip irrigation of cotton. Working with local farmers, the district is giving drip a tough test by comparing its performance to that of the most water-efficient sprinkler design now on the market—the low-energy precision application (LEPA) sprinkler. After the first year of trials, drip produced 19 percent more cotton per hectare than the LEPA-irrigated fields.[13]

With its long growing season, sugarcane is one of the thirstiest crops of all. Worldwide, the annual production of sugarcane consumes about as much water (including rainfall and irrigation) as all of the world's fruits and vegetables combined. Especially in dry regions, large-scale sugarcane production can consume a huge share of available water supplies. Sugarcane growers in the Indian state of Maharashtra, for example, take 50 percent of available irrigation water even though they occupy only 10 percent of the cropland. Nationwide, some 3.3 million hectares of India's cropland are planted in sugarcane, and more than 80 percent of this area is irrigated. A shift from conventional surface irrigation to drip on this 2.6 million hectares could potentially save substantial quantities of water in some of India's most water-needy regions.[14]

As with cotton, sugarcane often gets higher yields when watered by drip methods. Field studies by university researchers in Maharashtra, India, have found that drip-irrigated sugarcane yielded 12–30 percent more per hectare than sugarcane grown under conventional surface methods, while cutting water use by 30–65 percent. Similarly, studies in Hawaii have found that sugarcane grown under drip irrigation yielded 24 percent more and used half as much water as cane grown under other methods. With consistent findings of a two-and-a-half- to threefold jump in water productivity, conversion of irrigated sugarcane to drip methods could free up a large quantity of water for other crops or for urban uses—with no sacrifice in output.[15]

The third and probably most important driver behind the potential boom in drip irrigation is the recent development of more affordable drip systems. To date, with capital costs up to $2,500 per hectare, drip irrigation has mainly been used by wealthier farmers on high-value fruit and vegetable crops. In recent years, however, Denver-based International Development Enterprises (IDE) has designed and field-tested several different drip systems that are more affordable and appropriate for small farms.[16]

Conventional drip systems are expensive mainly because of the costs of the large amount of plastic tubing required, the emitters in the drip lines, and the filter needed to prevent clogging. IDE designed a system that costs one quarter as much as the conventional variety by making the drip lines portable, thereby allowing each line to serve 10 rows of crops instead of 1, by replacing expensive emitters with simple holes punched into the drip line, and by using off-the-shelf plastic containers and cloth filters in place of costly filtration equipment. The result is a system that costs just $50 per half-acre, or $250 per hectare. Field tests on vegetables in the hill areas of Nepal and on mulberry in Andhra Pradesh, India, showed that the system doubled the area under cultivation with the same volume of water.[17]

After these promising field trials, IDE began marketing low-cost drip in other parts of India. A number of these systems are now in place, for example, in the lower Himalayas of Himachel Pradesh, bordering Tibet. These hill farmers typically store water in either communal or private tanks that are supplied by springs or mountain runoff. During the dry months from March to June, they often run short of water and must leave a portion of their cropland fallow. When I visited this region in January 1998, farmers consistently mentioned one reason for their interest in the low-cost drip system: it would save enough water to allow them to double their cultivated area in summer. Few if any of them had factored into their equations the higher yields that were also likely. With New Delhi markets only five hours away by truck or train, the local farm economy in this region is poised for an infusion of cash income from the increased production. As of October 1998, IDE had helped install some 880 low-cost microirrigation (drip and sprinkler) systems in India and 470 in Nepal.[18]

IDE is now in the process of partnering with other institutions to launch a large-scale initiative to spread drip irrigation much more widely in developing countries. An early phase of the effort will involve demonstration projects in a number of key regions so that drip systems can be custom-built to suit local conditions and to ensure that feedback from farmers informs each system's design. One project, a joint effort of IDE and the World Bank, will test the potential of drip to save water at the tail-end of a canal system in India as a partial solution to the problems of water shortage and reduced crop production so common among tail-ender farmers. (See Chapter 10.)[19]

These lower-cost drip systems may set the stage for a quiet revolution in irrigated agriculture. For years, experts have acknowledged drip to be the most efficient form of irrigation, but have relegated it to a minor role globally because of its high cost and limited applicability. That role is now likely to expand. Made affordable and marketed effectively, drip has the potential to stretch scarce water supplies in north China, the cotton fields of the Aral Sea basin, the Indian subcontinent, much of the Middle East, parts of sub-Saharan Africa, and drought-prone northeastern Brazil, as well as in parts of Australia, southern Europe, and the United States.

Enter the Information Age

Along with technologies like drip that deliver water more effectively to crops, improved management practices can help farmers reduce their water demands while maintaining or increasing crop yields. Among the most exciting and potentially beneficial is the use of weather monitoring and satellite technologies to help farmers know when their crops actually need water. This information allows them to irrigate only when necessary and to apply only as much water as their crops need.

Many farmers still use the old "look and feel" method of determining the amount of moisture in the soil. An experienced irrigator can do fairly well with this approach, but in most cases it is not very accurate. In recent years, researchers and entrepreneurs have designed new weather monitoring and information systems that enable farmers to determine much more precisely when and how much to irrigate.

California's Department of Water Resources, for instance, operates a network of more than 100 automated and computerized weather stations in key agricultural areas of the state. Each station hourly collects local climate data, including solar radiation, wind speed and direction, relative humidity, rainfall, and air and soil temperature. These data are transmitted to a central computer in Sacramento, where they are checked for accuracy. For each station site, the computer then calculates the potential or reference evapotranspiration rate (ETo), which is the amount of water grass would transpire at that particular location. With this benchmark, farmers can calculate the evapotranspiration rate of their crops by multiplying the ETo by a factor specific to the particular crop. By knowing how much moisture is stored in the soil and how much the crops have consumed, a grower can determine when it is time to irrigate.[20]

Known as CIMIS, for California Irrigation Management Information Service, this operation is the largest automated agricultural weather station network in the United States. Some of the stations are owned and operated by the state; others, by local agencies. A 1995 survey conducted by the University of California at Berkeley found that, on average, farmers using CIMIS to schedule their irrigations experienced an 8-percent yield increase and a 13-percent reduction in water use.

Economic benefits ranged from $99 per hectare of alfalfa to $927 per hectare of lettuce.[21]

Similar statewide agricultural weather networks in the United States include the Washington State University Public Agriculture Weather Systems (PAWS) and the Arizona Meteorological Network (AZMET), which generates weather data 21 hours a day and which farmers can get to free via a computer bulletin board. A more sophisticated, quasi-commercial service called the Northwest Irrigation Network operates in the Pacific Northwest. It consists of a weather monitoring network and a satellite information system run cooperatively by the U.S. Bureau of Reclamation, farming communities, and an Oregon-based firm called IRZ Consulting.[22]

As with the statewide networks, the Pacific Northwest system includes stand-alone remote stations that collect weather data in agricultural areas. These stations then beam the data to a communications satellite, which in turn transmits the information to computers at the U.S. Bureau of Reclamation office in Boise, Idaho. At midnight each day, the Boise computers send the data to IRZ's computers in Oregon. There, a computer program analyzes the data relative to the known field and crop conditions of IRZ's farmer clients. By telephone, fax, or the Internet, the company lets growers know how much water to apply to their crops. IRZ, which now serves more than 100,000 hectares of farmland in northeast Oregon and southern Washington, also offers a menu of other irrigation management services to supplement the customized irrigation scheduling advice. According to IRZ president Fred Ziari, farmers using the service typically average water savings of 15–20 percent.[23]

Farmers who own computers can also time their irrigations more accurately by using one of several popular computer programs. The USDA's Natural Resource Conservation Service (NRCS) offers a program called *NRCS Scheduler* that is applicable to most climates and a wide variety of crops. The farmer enters data on field conditions, local weather, and the crops being grown, and the program calculates daily and monthly evapotranspiration, soil moisture changes, and irrigation requirements throughout the growing season. It can also generate end-of-season reports that let the farmer know whether too much or too little water was applied.[24]

At the global level, the International Water Management Institute in Sri Lanka has developed a computerized tool for irrigation and crop planning called the *World Water and Climate Atlas*. The atlas, available on CD-ROM and through the Internet, integrates agricultural weather data from 56,000 weather stations around the world for the period 1961 to 1990. Farmers, agronomists, irrigation planners, and others can use the long-term climate data to better match crops and crop varieties to local climate conditions. It also delineates areas where moisture is adequate to grow certain crops, as well as those where irrigation is necessary. The electronic atlas could be used, for example, to pinpoint areas where crops now grown under rain-fed conditions could benefit from some supplemental irrigation to safeguard against water shortage or drought. Because it covers all regions, the atlas can be used to aid crop planning and water management in rich and poor countries alike.[25]

Finally, researchers at the University of Texas at Austin have developed a digital atlas of the world's water balance that allows users to calculate monthly soil-water balances. Using a mesh of cells 50 kilometers square that covers the entire Earth, the atlas helps trace the movement of water between the atmosphere, soil, rivers, and groundwaters. Accessible through the Internet, it can be used to predict soil moisture, evapotranspiration, and water deficits, and in this way offers another aid to irrigation planning.[26]

Programs That Work

A successful strategy to increase irrigation efficiency and water productivity almost always involves a mix of technical, managerial, and institutional measures that reinforce each other and that are appropriate to the local farming culture and the crops being grown. How water is priced and allocated, whether or not farmers can sell their water, and whether or not groundwater pumping is regulated are among the factors influencing farmers' decisions. Changing these "rules of the game," as discussed further in Chapter 10, can be instrumental in encouraging more-effective water use.

Irrigators in California's Broadview Water District, for example, have shown that pricing water more effectively can lead farmers to save water. Located on the west side of the San Joaquin valley, an area with salty and contaminated farm drainage (see Chapter 5), the water district had been pricing water like many districts do—at a flat rate. Farmers paid $16 per acre-foot (1.3¢ per cubic meter), regardless of the volume of water used. In the late 1980s, faced with the need to reduce drainage into the highly polluted San Joaquin River, the district replaced this system with a two-tiered structure. For each crop, the district determined the average volume of water used over the 1986–88 period, and then applied the same base rate of $16 per acre-foot to 90 percent of this amount. Any deliveries made above that level were charged at a rate of $40 per acre-foot—2.5 times higher, reflecting the cost to the district of collecting and discharging the surplus drainage.[27]

In addition to the pricing incentives, the district offered farmers greater flexibility in water delivery schedules, provided them with water use information specific to their fields, and improved some of its own operations. The key incentive, however, was the change in pricing. Even though farmers were still paying much less than the real cost of their irrigation water, they had sufficient incentive and available ways to boost irrigation efficiency. On average, cotton growers used 25 percent less water over the period 1990 to 1993 compared with 1986–89. Similarly, water use on tomatoes fell by 9 percent, on cantaloupes by 10 percent, on wheat by 29 percent, and on alfalfa seed by 31 percent. Crop yields either held steady or increased. Drainage releases from the district fell by 36 percent because of the reductions in water use and the capture and recycling of water that would otherwise have run off the fields.[28]

There are very few examples of whole countries, river basins, or large farming regions that have orchestrated the kind of water productivity gains that will be needed during the twenty-first century. The handful that do exist, however, are instructive.

Not long after Israel's founding, many of its best scientists began working on a fundamental problem: the young nation's water resources were not sufficient to irrigate all of its irrigable land. The challenge was to stretch the water supply as much as possible while still maintaining a productive and viable farming economy. The Water Commission, situated in the Ministry of Agriculture, adopted a variety of measures to drive Israel toward greater irrigation water productivity.

First, the commission established fixed allocations of water to farms based on the area cultivated, the crops grown, and the water requirements of those crops. Under such a system, farmers irrigating inefficiently would suffer, because they typically would be unable to irrigate their whole farm. Second, irrigation water was priced according to an increasing block rate structure, similar to the tiered structure of the California water district just described. Third, the government passed the Law of Water Metering, which stated that anyone producing, supplying, or consuming water had to measure it. This ensured that trends in water use could be monitored accurately, and that water users could be charged according to the volume actually used.[29]

Early on, Israel's Water Commission also established an

Irrigation Efficiency Unit to oversee research and development specifically aimed at boosting crop yield per unit of water. The group ran pilot projects and large-scale demonstration farms to test various methods in the field. A special agency called the Irrigation and Soil Field Service also was established in the 1950s to focus exclusively on extension services—helping farmers select the best practices and methods for their conditions. And finally, the Israeli government offered low-interest loans to farmers wishing to install more-efficient irrigation systems.[30]

All in all, this multipronged strategy produced striking results: Between 1951 and 1985, Israel expanded its irrigated area fivefold with only a threefold increase in water use. Nationwide, water use per hectare dropped from an annual average of 8,200 cubic meters in 1951 to about 5,200 cubic meters in 1985—a decline of 37 percent. Over the same period, output per cubic meter nearly tripled and the value of output (in real terms) jumped 10-fold. Israel is the only nation that appears to have done what the world needs to do over the next 30–40 years—double water productivity in agriculture.[31]

Impressive as the Israeli story is, the gains were mainly achieved with vegetables, fruits, and other high-value crops. Meeting a global challenge of doubling water productivity will require equally impressive gains with wheat, corn, rice, and other grains—the staples of the human diet and the most widely irrigated crops. Although no country has even come close to showing how to do this, a farming region in the Texas panhandle offers some inspiration and insights.

In dry and drought-prone northwest Texas, farmers depend on the dwindling Ogallala aquifer for their irrigation water. (See Chapter 4.) During the 1980s, with various studies warning about large declines in irrigated area and crop production as the region's water tables fell, local water officials and researchers joined forces to stem the pending economic disruption. They created and pushed a package of technologies and management practices that, by lifting water productivity, has kept many irrigators in business and slowed the Ogallala's depletion.

Spearheaded by the High Plains Underground Water Conservation District in Lubbock, which overseas water management in 15 counties of northwest Texas, the effort has involved a major upgrade of the region's irrigation systems. First, many conventional furrow systems, with typical efficiencies of 60 percent, have been equipped with surge valves that raise irrigation efficiency to about 80 percent. Just as the name implies, surge irrigation involves sending water down the furrows in a series of pulses or surges rather than in a continuous stream. The initial pulse somewhat seals the soil, letting subsequent surges flow more quickly and uniformly down the field. This evens out the distribution of water, allowing farmers to apply less at the head of their fields while still ensuring that enough water reaches crops at the tail-end.[32]

A time-controlled valve alternates the flow of water between rows, and can be set at different cycle lengths and flow rates depending on the soil type, the length of the furrow, and other conditions. When combined with soil moisture monitoring and proper irrigation scheduling, surge can cut water use by 10–40 percent compared with conventional gravity irrigation. Savings in the Texas High Plains have averaged about 25 percent, and within two years farmers there have generally recouped their initial investment, which ranges from $30 to $120 per hectare depending on whether piping is already in place.[33]

Many farmers in the region have also upgraded to efficient sprinklers. Especially in dry, windy areas like the U.S. Great Plains, spraying water high into the air can lead to large water losses from evaporation and wind drift, negating the usual efficiency gains of sprinklers. The High Plains water district has encouraged farmers to adopt one of two varieties of low-pressure sprinklers. One type delivers a light spray from nozzles about a meter above the soil surface, and typically registers efficiencies of 80 percent, about the same as surge irrigation. (See Table 8–3.)

A second variety, however, does substantially better. Low-energy precision application sprinklers deliver water in small doses through nozzles positioned just above the soil surface. They nearly eliminate losses from evaporation and wind drift, and typically raise efficiency to 95 percent. Under LEPA irrigation, corn yields have increased about 10 percent, and cotton yields by 15 percent. Combined with 15–35 percent water savings over other methods, these yield increases lead to substantial gains in water productivity. Farmers converting to LEPA typically recoup their investment in two to seven years, depending on whether they are upgrading an existing sprinkler or purchasing a new one. Virtually all the sprinklers in the 15-county water district area are now either the low-pressure spray or LEPA variety.[34]

Along with research and development, the Texas High

Table 8–3. Efficiencies of Selected Irrigation Methods, Texas High Plains

Irrigation Method	Typical Efficiency	Water Application Needed to Add 100 Millimeters to Root Zone	Water Savings Over Conventional Furrow[1]
	(percent)	(millimeters)	(percent)
Conventional Furrow	60	167	—
Furrow with Surge Valve	80	125	25
Low-Pressure Sprinkler	80	125	25
LEPA Sprinkler	90–95	105	37
Drip	95	105	37

[1]Data do not specify what portion of savings result from reduced evaporation versus runoff and seepage.
SOURCE: Based on High Plains Underground Water Conservation District (Lubbock, Texas), *The Cross Section*, various issues.

Plains program has also included substantial extension work to help farmers adopt water-saving practices. Extension spread the word about furrow diking, for example, one of the most economically beneficial measures many farmers can adopt. Farmers form small earthen ridges across furrows at regular intervals down the field. The small basins trap both rainwater and irrigation water, promoting more infiltration into the soil. Furrow diking is a key to obtaining the highest possible irrigation efficiency with LEPA, for example, and to storing as much pre-season rainfall in the soil as possible.

Constructing furrow dikes costs about $10 per hectare. James Jonish, an economist at Texas Tech University, points out that if furrow dikes capture an extra 5 centimeters of rainfall in the soil, they can boost cotton yields by up to 225 kilograms of lint per hectare, a potential economic gain of $400 per hectare, depending on cotton prices. In contrast, getting those higher yields by pumping 5 centimeters of additional groundwater would cost $15–22 per hectare, 50–120 percent more, and would hasten the aquifer's depletion.[35]

Overall, the Texas water district's program, which has been bolstered by state-funded low-interest loans for efficiency improvements, has achieved impressive results. Over the last two decades, growers have boosted the water productivity of cotton, which accounts for about half the cropland area, by 75 percent. Fully irrigating an acre of land used to require a groundwater well able to produce at least 10 gallons a minute per acre, but the district now considers 2–3 gallons a minute sufficient, with no negative effect on yields. Between spring 1997 and spring 1998, the average district-wide depth to the water table fell 4 inches, compared with an average annual drop for the previous decade of 1 foot. Although more rainfall contributed to the reduced groundwater pumping in this most recent year, district officials credit the growers' conservation efforts as well.[36]

Despite these successes, High Plains irrigated agriculture

must continue to improve its performance to stay viable. The district is now ramping up a program to bring the most up-to-date knowledge about crops and water use together with computerized information systems and water-efficient technologies, all aimed at raising water productivity even further.

Using weather and crop water use data, the district develops irrigation schedules based on a two-and-a-half-day cycle of LEPA irrigation, rather than the usual five- to seven-day cycle. The principles at work are much the same as with drip irrigation—delivering small volumes of water at frequent intervals to create a near-ideal moisture environment for the plants. Preliminary results with corn and cotton show promising yield increases. The district is developing a computer program that integrates real-time weather data with this method of high-frequency irrigation. Water district assistant manager Ken Carver expects the program to be available in 1999 and to get widespread use.[37]

What is so exciting and promising about the developments in northwest Texas is that they offer a way to irrigate corn, wheat, and other grains nearly as efficiently as drip systems irrigate fruits, vegetables, and cotton. And like drip, these systems offer a yield bonus as well. Efforts like these make the goal of doubling global water productivity seem more a possibility than a pipe dream.

Rice—A Tough Challenge

About 100 million of the world's 255 million hectares of irrigated land are planted in rice, the preferred staple for about half the human population. More than 90 percent of the world's rice is produced in Asia, where many major rivers are now tapped out during the dry season and where city-farm competition for water is escalating. Finding ways of growing rice with less water is critical to sustaining the harvest of this important crop, much less expanding it.[38]

Producing rice takes a lot of water not because the plant uses water inefficiently, but because of how and where rice is grown. Farmers typically start rice seedlings in nursery plots, and then transplant them to paddy fields. To control weeds, they flood the paddies prior to transplanting and keep them flooded for most of the growing season. The rice plant is unique in that its roots do not need to take in oxygen from air pockets in the soil, so the plant can thrive in waterlogged conditions. Flooding rice fields, however, can result in substantial evaporation losses, especially until the crop cover is established. In northeast Sri Lanka, for example, researchers found that evaporation accounted for 29 percent of total dry-season water consumption from rice fields.[39]

Rice is also a thirsty crop because of where it is grown. The world's major rice belt stretches from eastern India eastward through Indonesia. In these hot climates, transpiration rates are high. As a result, rice consumes more water than an equivalent yield of wheat, which is grown in cooler climates.

Over the last quarter-century, the widespread adoption of high-yielding and early-maturing rice varieties boosted not only land productivity but water productivity as well. These new varieties contributed to a doubling of average rice yields and reduced the growing season for rice from 140 days to 110 days. Together, these gains brought about a 2.5- to 3.5-fold increase in the amount of rice harvested per unit of water consumed—a tremendous achievement. There is no strategy in sight that could lead to a repeat of those gains on a wide scale. But scientists have identified some possible new ways to grow rice with less water.[40]

Among the most promising options is to sow rice seeds directly in the field rather than transplanting rice seedlings. Motivated mainly by rising labor costs, farmers in a number of Southeast Asian countries have tried this method. In one irrigation system in central Luzon in the Philippines, for instance, farmers who shifted to the direct seeding method used less water during both the preplanting and growing periods, and ended up with a 9-percent increase in yield as well. Overall, water productivity increased 25 percent. Researchers are quick to point out that they do not have enough data and information to know how widely applicable direct seeding may be, and that greater competition from weeds may be a problem with this method. But direct seeding of rice is clearly worthy of more attention.[41]

In Malaysia, the Muda irrigation scheme achieved very impressive results by combining direct seeding with improved water monitoring and management. During the 1980s, Muda rice farmers and irrigation managers were faced with a drop in rainfall and a simultaneous decline in the volume of water available for irrigation, jeopardizing the area's rice production. They responded with a number of changes. First, many farmers shifted to direct sowing of seeds in the field, reducing water use and costly labor as well. For its part, the Muda Agricultural Development Agency shored up tertiary canals to improve the distribution of water throughout the irrigation project. The agency also put in place a computerized data gathering and feedback system that tracked the depth of water in the Muda rice paddies. This allowed them to optimize the use of rainfall, and to release irrigation water only when necessary.[42]

Bolstered by supportive government rice policies, this combination of measures had a markedly positive effect. Despite the decline in water availability, Muda rice growers increased total crop output by about 16 percent between the early 1980s and early 1990s. Water productivity—measured as paddy output per unit of water released from the scheme's reservoirs—rose 45 percent. Whereas the Muda scheme accounted for one quarter of Malaysia's total paddy production in 1981, it accounted for half the harvest 10 years later.[43]

In contrast to wheat and corn, virtually none of the world's rice is irrigated by sprinklers. Because sprinklers typically apply water to a field more uniformly than gravity methods do, they allow farmers to irrigate their crops adequately with less water. In the United States, where rice is seeded directly rather than transplanted, researchers have found that sprinkler irrigation may substitute for flood irrigation of some rice varieties. In Arkansas, they found that irrigating rice with sprinklers cut water use by half and gave yields comparable to flood irrigation. Results were more mixed in Louisiana and Texas, with sprinkler-irrigated rice yielding 10–25 percent less than flooded rice. Since this method represents an entirely new crop management regime, it seems worthy of further research, design, and testing—especially with the low-cost sprinklers appropriate for smallholder plots that are now hitting the market. (See Chapter 9.)[44]

Water-Thrifty Crops

Most of the options for raising agricultural water productivity discussed so far have to do with knowing when and how much irrigation water is needed and then delivering that water more efficiently to the roots of crops. Another set of options has to do with improving the ability of crops themselves to use water more efficiently—to yield more food for every liter of water available.

Any plant's growth is inextricably linked to its rate of transpiration, the process by which it takes in moisture and releases it back to the atmosphere. Working with a variety of crops in a variety of locations, researchers have consistently found that there is a linear relationship between a crop's transpiration and its yield of plant matter, up to the point at which water is no longer a limiting factor. This means that for any particular crop in a given location, more plant production requires more water consumption.

Because of this fundamental relationship between water and crops, agronomic options for getting "more crop per drop" are inherently limited. Moreover, scientists have already exploited some of the most fruitful possibilities. For example, increasing the proportion of a plant's total biomass that is harvestable yield—known as the harvest index—is one way to get more edible crop from the same amount of water. The hybrid wheat and rice varieties that spawned the Green Revolution, for example, were bred to allocate more of the plant's energy—and thus water use—into edible grain. This greatly boosted both the land and water productivity of the major irrigated grains. However, most plant breeders see little scope for further raising the harvest index of wheat and rice, now around 50 percent.[45]

Another possibility is to breed or bioengineer plants that can photosynthesize in a more water-efficient manner. So-called C_4 crops, which include corn, for example, have special anatomical and biochemical properties that allow them to photosynthesize more efficiently, increasing water use efficiency as well. But incorporating these features into other crop types without sacrificing other valuable traits is problematic. Bioengineers are also examining possibilities for altering plants' stomata—the leaf openings through which plants exchange gases with the atmosphere. Because the stomata stay open longer than necessary, plants release some water to the atmosphere without getting any growth gain in return. Getting the stomata to close more quickly might reduce plant water use without sacrificing any yield. Some researchers hold out hope for agronomy's holy grail—altering photosynthesis itself. But most people view this as a long shot because the process has remained essentially unchanged for some 2 billion years of plant evolution and natural selection.[46]

In recent years, advances in genetic engineering have opened up possibilities for more efficiently building traits such as drought resistance and salt tolerance into crop varieties, which can safeguard crop yields under hostile environmental conditions. In particular, a tool known as molecular marking has enabled scientists to more easily find genes that harbor desirable traits. The genome of wheat is five times larger than the human genome, so finding any specific gene unaided would be extremely difficult. Crop geneticist Pamela Ronald of the University of California at Davis likens the task to finding a house in New York City or Tokyo without knowing the address.[47]

With new methods of molecular marking, however, scientists can create a sort of map or guide to the crop's genome, helping locate particular genes. Ronald and her colleagues, for example, have used these techniques to insert genes for disease resistance into rice, which has a genome about one seventh as large as the human genome and just 3 percent as large as wheat's. Similarly, scientists may be able to incorporate genes for salt tolerance or drought resistance into modern crop varieties. Such changes may lower the crop's yield potential, but could still benefit many farmers. In drought-prone areas, many growers are willing to accept lower average crop yields in exchange for added insurance against drought-induced catastrophic crop losses.[48]

Shifting the location and timing of crop production also presents opportunities for getting more crop per drop. Plants transpire at a rate determined by the difference in vapor pressure between their stomata and the surrounding atmosphere. In a dry season or climate, the vapor pressure deficit between a crop and the air around it is large, and so moisture will flow more rapidly through the plant's stomata into the atmosphere.

Theoretically, shifting a larger portion of global crop production to more humid climates, cooler climates, or, within a given climate, to cooler and more humid times of the year can save water because it would reduce crop transpiration. However, gains in one area often come at the expense of another. For instance, humid regions would tend to have more cloud cover, which would reduce the influx of sunlight, another key driver of plant growth. For a crop to perform well in cooler weather it may need to have special physical or biochemical traits that take energy away from the plant's growth or that lower its harvest index. Big gains from such shifts thus seem unlikely.

There may also be untapped potential to improve and spread the cultivation of traditional crop varieties, which can build both diversity and resilience into regional food production systems. Many grains native to Africa, for instance, are naturally hardier, more heat- and drought-tolerant, and need less irrigation water than modern varieties do. African rice, for one, matures quickly, and therefore has comparatively low water requirements. Fonio, one of the world's best-tasting grains and native to West Africa, is among the world's fastest-maturing cereals. These traditional crops have lower average yields than modern varieties but may nonetheless have a greater role to play as drought, water scarcity, and increasing competition for water further impinge on agriculture.[49]

Designing for More Than One Use

In many ways, the last set of options for improving water productivity is the most obvious—using water more than once. A great deal of water recycling and reuse happens naturally as the water used by a city, farm, or factory returns to a river or an underground aquifer and gets picked up again by another user downstream. Many irrigators in South Asia, for example, pump groundwater that was first used by a farmer supplied by a public canal system. Seepage from unlined canals and excess water from the first farmer's field recharge the underlying aquifer, allowing farmers with groundwater wells to tap it again for irrigation.

Because crops do not require water that meets drinking-quality standards, many options for irrigating with lower-quality water exist. A number of countries are stretching water supplies, for instance, by reusing municipal wastewater for irrigation. With urban water use in many developing countries likely to at least double over the next three decades, treated wastewater will be both an expanding and fairly reliable supply.

Many large coastal cities currently dump their wastewater, treated or untreated, into the ocean, not only wasting water but also polluting and degrading the marine environment. As long as wastewater is kept free of heavy metals, harmful chemicals, and dangerous levels of disease vectors, it can become a vital new supply for agriculture.

Israel now reuses 65 percent of its domestic wastewater for crop production. Treated wastewater accounts for 30 percent of the nation's agricultural water supply, and this figure is expected to reach 80 percent by 2025. Tunisia irrigates a modest 3,000 hectares with treated wastewater, but plans to increase this area 10-fold. And Tenerife, the largest of the Canary Islands off the coast of northwest Africa, is coping with limited groundwater sources by treating and reusing wastewater for irrigation of bananas, its main export crop.[50]

Farmers and irrigation districts can also recycle and reuse their own irrigation water. Although much of this happens naturally, more can be done with planning, good design, and proper investments. In Malaysia's Muda rice irrigation scheme, water managers installed six pumping stations to capture and recycle the irrigation drainage flowing out of the system. In this way, farmers in the scheme get an additional 123 million cubic meters of water a year, increasing the supply from the project's reservoirs by 17 percent. Similarly, recycling of drainage water in a rice system in Niigata Prefecture in Japan boosted the scheme's water supply by about 15 percent. Recycling and reuse always have downstream consequences—whether positive or negative—so these strategies need careful evaluation.[51]

Finally, as agriculture comes under increasing pressure to return a portion of its water supplies to restore the health of rivers, wetlands, and other freshwater ecosystems, there may be creative opportunities for irrigation water to provide ecological benefits as well. I got a glimpse of these possibilities when traveling through the delta of the Colorado River in northern

Mexico in 1996. This area was once a lush land of great biological richness, supporting some 200–400 plant species and numerous birds, fish, and mammals. American naturalist Aldo Leopold, who canoed through the delta in 1922, called it a "milk-and-honey wilderness" and a land of "a hundred green lagoons."[52]

Today the delta is a desiccated place of mud flats, salt flats, dry sand, and scattered murky pools. The treaties that divide up the river among Mexico and seven states in the United States allocate more water to the eight parties than the river actually carries in an average year. As a result, virtually no water flows through the delta and out to sea, which has decimated the delta environment.

During my travels, however, I visited a place that resembled what the delta must have looked like before big dams and river diversions siphoned off the Colorado's flow. The Ciénaga de Santa Clara is a contiguous wetland covering some 20,000 hectares. In an otherwise desolate landscape, the Ciénaga looks like a lush green jewel. It is a major stopover point for migratory birds along the Pacific Flyway, and may be home to the largest remaining populations of endangered Yuma clapper rails and desert pupfish.[53]

What most astonished me about the Ciénaga, however, was its source of water. This critical wetland is the fortuitous and unplanned consequence of an agricultural drainage canal that extends 80 kilometers from the Wellton-Mohawk irrigation district in Arizona to this southern portion of the delta. The U.S. government built the canal in 1977 to redirect the irrigation drainage away from the Colorado River so as to reduce the loading of salts and agricultural chemicals just before the river crossed into Mexico. The drain was supposed to be a temporary solution to the water-quality problem until the government could build and begin operating a large desalting plant at Yuma, Arizona. Because the government never put the desalting plant into full operation (although the idea of treating and selling the drainage has been floated in recent years), the Wellton-Mohawk drainage has continued to sustain the Ciénaga.[54]

The Ciénaga's unplanned existence and biological importance suggest that there may be many unrealized possibilities for irrigation water to produce environmental benefits. Although poor in quality, the Wellton-Mohawk drainage has created and sustained one of the largest desert wetlands in the American Southwest and helped prevent at least two species from becoming extinct. Ideally, the United States and Mexico would agree to dedicate freshwater flows to restore a portion of the delta environment. But in the absence of such an agreement, this irrigation drainage is playing a critical role. California's tragic experience with selenium-laced drainage entering the Kesterson wetlands (see Chapter 5) is a reminder that this kind of arrangement needs to be carefully planned and monitored. Done right, however, the reuse of irrigation water to expand wetland habitat is a way of increasing water productivity—getting more value out of each liter we remove from rivers, lakes, and aquifers.

Another example of dual benefits and multiple use of water is taking shape in California. A unique alliance of rice farmers and environmentalists is turning off-season rice fields into useful waterfowl habitat. California's Central Valley, now known for its multibillion-dollar agricultural industry, once supported among the largest concentrations of migratory waterfowl on Earth. As recently as 1950, some 60 million birds spent winters in the valley. Today, with the state having lost 95 percent of its wetlands, the migratory bird population has fallen to 3 million.[55]

Sacramento valley rice farmers typically flood their fields from April through mid-October, the growing season for California rice. After the harvest, they drain the fields and leave them dry until the next spring's planting. In the past, they got rid of the tough stubble left behind after harvest by burning it. But following an especially bad pollution episode that darkened the skies of the state capital, legislators decided in 1991 to gradually phase out the practice of burning rice straw.[56]

Without a ready market for the hardy straw, one of the few alternatives available was for California growers to flood their fields during the fall and winter months to promote the stubble's decomposition. In much the same way the disposal of irrigation drainage to the Colorado delta created a rich new wetland, the fall flooding of rice fields created an environmental boon as well—new habitat for migrating waterfowl. To geese, cranes, and ducks, a flooded rice field is just about as enticing as a natural wetland. The birds feed on the stubble, the leftover rice, and the midges, worms, crayfish and other small animals that inhabit the flooded fields. In return, the birds help break down the rice stubble and leave natural fertilizer behind.

Sensing win-win possibilities, rice growers and environmentalists formed the Ricelands Habitat Partnership to explore options for helping the rice industry manage its straw problem while at the same time turning parts of the Central Valley back into prime waterfowl habitat. One hurdle for the initiative is that the additional water demands of turning ricelands into waterfowl habitat may complicate efforts to restore and protect salmon runs—a major state environmental goal. Farmers would be drawing more water in the fall, when river flows are already low and salmon are migrating. The partnership is now talking to water districts, the fishing industry, and other members of the environmental community to explore solutions that take all identified concerns into account.[57]

———

As this chapter has described, by thinking creatively and systemically—whether on a farm, within an irrigation district, or more broadly within a region or river basin—numerous possibilities exist for increasing the productivity of agriculture's use of water. Many of these measures will take more research and development. Many involve complex management challenges that will test institutions and shake up the status quo. Together, however, these strategies and measures allow us to at least envision a modern irrigation-based society that is more efficient, productive, ecologically sound, and potentially lasting.

9

THINKING BIG
ABOUT SMALL-SCALE
IRRIGATION

During the early months of any year, the vast plains of eastern India reveal a curious piece in the puzzle of global agriculture. Field after field appears barren and dry, even though vast quantities of water lie a short distance below the surface. The families who farm these plots rank among the world's hungriest and poorest, so they desperately need to coax more food from the earth. The benefits of irrigation, however, have eluded them. They have no way to tap the plentiful water beneath their land because the available technologies are too big and too expensive for their small plots.

These farmers in eastern India are not alone. There are tens of millions of others, particularly in South Asia and sub-Saharan Africa, who have missed out on the benefits of irrigation. About three quarters of the farmers in developing countries cultivate less than 2 hectares of land. Some have access to canal water from large government-subsidized systems, but many do not. The less fortunate usually earn no more than $300 a year, which places them among the world's acute poor—the 1.3 billion people who survive on less than a dollar a day. Typically, the cheapest way to irrigate with groundwater is to install a diesel pump on a tubewell, which costs at least $350—way out of reach for these small farmers.[1]

For all its successes, the modern irrigation age has a deep structural flaw: it has bypassed the majority of the world's farmers. Partly because so many farmers lack access to irrigation, poverty and hunger remain entrenched in large pockets of the countryside. Like trickle-down economics, trickle-down food security does not work well for the very poor. India, for example, is self-sufficient in grain, yet tens of millions of Indians suffer from hunger and malnutrition. Globally, the grain harvest is more than adequate to feed everyone a sufficient diet, yet one out of seven people in the world does not get enough to eat.[2]

The vast majority of the world's 840 million undernourished people do not participate in the integrated global economy we hear so much talk about today. The production and trade of more "surplus" food will not solve their problems of hunger because many of them are too poor to buy that food, even at today's historically low prices. The surest and most direct way of reducing hunger among the rural poor is to raise their productive capacities directly. Access to irrigation, in turn, is one of the surest ways of boosting small-farm productivity. With a secure water supply, farmers can invest in higher-yielding seeds and harvest an additional crop or two each year. Irrigated plots in developing countries commonly yield twice as much as rain-fed plots. (See Figure 9–1.)[3]

Many of the world's poorest farmers will never benefit from large public irrigation schemes. As discussed in Chapter 3, high costs and concerns about social and environmental sustainability have slowed the spread of large irrigation projects. Big government schemes, moreover, rarely benefit the poorest of the poor. Raising the production potential of small, poor farm-

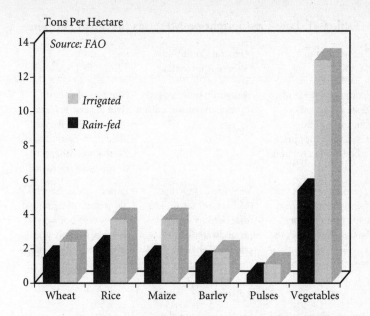

Figure 9–1. Average Yields on Rain-fed and Irrigated Land, Developing Countries

ers requires irrigation technologies specifically geared to their farm sizes and income levels.

A spectrum of irrigation technologies—custom-designed for small plots and affordable by the poorest farmers—has begun to open the door to irrigation's benefits for millions of farm families, from the Gangetic plains of South Asia to the terraced hills of Nepal to the drylands of sub-Saharan Africa. (See Table 9–1.) Collectively, these little-known technologies could lift productivity on tens of millions of hectares of farmland. The enhanced food security and rising incomes of large numbers of small-scale farm families can be a powerful engine of economic growth in the world's poorest regions and help stem the exodus of rural dwellers to burgeoning cities. But that engine must be jump-started. Especially as large-scale water

Table 9–1. Low-Cost Irrigation Methods for Small Farmers

Technology or Method	General Conditions Where Appropriate	Example Locations
Cultivated wetlands, delta lands, valley bottoms; flood-recession cropping; rising-flood cropping	Seasonally waterlogged floodplains or wetlands	Niger and Senegal river valleys; *fadama* of northern Nigeria; *dambos* of Zambia and Zimbabwe; other parts of sub-Saharan Africa
Treadle pump, rower pump; pedal pump; rope pump; swing basket; Archimedian screw; *shadouf* or beam and bucket; hand pump	Very small (less than 0.5 hectare) farm plots underlain by shallow groundwater or near perennial shallow streams or canals in dry areas or areas with a distinct dry season	Eastern India; Bangladesh; parts of southeast Asia; valley bottoms, *dambos, fadama,* and other wetlands of sub-Saharan Africa
Persian wheel; bullock and other animal-powered pumps; low-cost mechanical pumps	Similar to those above, but where average size of farm plots is roughly 0.5–2.0 hectares	Those above, in addition to parts of North Africa and Near East
Various forms of low-cost microirrigation, including bucket kits; drip systems; pitcher irrigation; as well as microsprinklers	Areas with perennial but scarce water supply; hilly, sloping, or terraced farmlands; tail-ends of canal systems; can apply to farms of various sizes, depending on the microirrigation technique	Much of northwest, central, and southern India; Nepal; Central Asia, China; Near East; dry parts of sub-Saharan Africa; dry parts of Latin America
Tanks; check dams; percolation ponds; terracing; bunding; mulching; other water harvesting techniques	Semiarid and/or drought-prone areas with no perennial water source	Much of semiarid South Asia, including parts of India, Pakistan, and Sri Lanka; much of sub-Saharan Africa; parts of China

SOURCE: Global Water Policy Project, Amherst, MA.

projects come under closer scrutiny for their environmental and social costs, a coordinated effort to provide access to small-scale irrigation methods that poor farmers will invest in and use is a critical addition to the portfolio of irrigation advances.

The Power of Treadle Pumps

Bangladesh, a Wisconsin-sized country of 125 million people, suffers from both too much water and too little. The monsoons bring devastating floods that often kill not only crops but people. Heavy rains are followed by a dry season during which crops can wither from lack of moisture. Without access to irrigation, many farmers simply do not bother to plant much during this dry period. Like their neighbors in eastern India, they leave their plots fallow, which in turn leaves them hungry and poor.

Over the past 16 years, however, large areas of Bangladesh have been transformed by a human-powered irrigation device called a treadle pump. To an affluent westerner, this manual pump resembles a Stairmaster exercise machine, and is operated much the same way. Grasping a horizontal arm on the bamboo frame to steady the body, the user pedals up and down on two long poles, or treadles. The pedaling motion activates a rope-and-pulley with two plungers, each positioned within a metal cylinder. A suction inlet at the bottom welds the twin cylinders together. On the upward pumping stroke, groundwater is sucked up into the cylinders while water from the previous stroke is expelled directly into a field channel. The volume of water that can be pumped in an hour depends on the distance to the water table, the diameter of the pump cylinders, and the energy expended by the operator. In general, the treadle pump becomes too inefficient and impractical if the groundwater is more than 6 meters below the surface.[4]

Designed by Gunnar Barnes, a Norwegian engineer, the treadle pump was introduced in Bangladesh in the early 1980s. Since then, the simple device has quietly revolutionized small-farmer agriculture in this poor South Asian country.

On an early morning in January 1998, I visited the fields of Brahmanbaria, about 80 kilometers northeast of Dhaka, and found them bustling with treadle pump operators—young and old, male and female. Colorful canopies decorated many of the pumps and provided relief from the hot sun. This is Bangladesh's dry season, and since there is no risk of flooding, farmers with treadle pumps can plant higher-yielding and higher-valued rice and vegetables with confidence that they will get a good harvest. Several farm families I talked with said that they had previously relied only on rainfall, and that since purchasing their treadle pump, they had doubled their crop production. With income generated from the first pump, some farmers had bought a second one. The pumps also gave them independence from local water lords, a more reliable supply of water, and precise control over when and how much they irrigated.[5]

Poor farmers are very risk-averse, and they typically invest only in technologies that will pay them back within one year. The treadle pump easily satisfies this criterion. Each pump irrigates about a fifth of a hectare (half an acre) and costs about $35, including installation of the tubewell. Net returns average more than $100 per pump, so farmers recoup their investment in less than a year, and often in one season.[6]

As anyone who has exercised on a Stairmaster knows, pedaling against pressure is hard work. Many families operate their treadle pumps four to six hours a day during the dry season. Adults treadle before sunrise and after they return from their daily labors; children treadle before and after school. This human labor has led some agricultural specialists to view the treadle pump as backward, even inhumane.

But it is hard to argue with the numbers. Farmers have purchased more than 1.2 million treadle pumps in Bangladesh alone. These have enabled families previously stalked by hunger to feed themselves year-round. With this simple technology, farmers are raising the productivity of more than a quarter-million hectares of farmland, and injecting an additional $325 million a year into the poorest parts of the Bangladeshi economy. The pumps contributed to the groundwater revolution that fueled the 136-percent rise in Bangladesh's irrigated area between 1975 and 1995, a period when global irrigated area rose by only 35 percent. (See Figure 9–2).[7]

To understand why the treadle pump has succeeded, and to grasp the full challenge of spreading small-scale irrigation in the developing world, I had to leave the rice and vegetable fields of Brahmanbaria and wind through the back roads of Dhaka to visit the small factories that make treadle pumps and the dealerships that sell them. The treadle pump's low cost, ease of repair, and simple but elegant design explain only a part of its success. Much has to do with its marketing—an activity sometimes maligned and often ignored by development groups and international agencies.

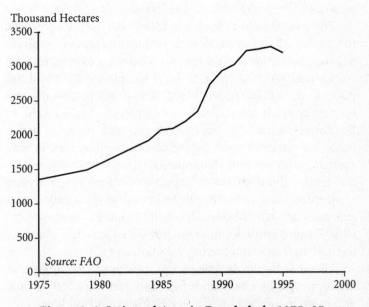

Figure 9–2. Irrigated Area in Bangladesh, 1975–95

Many technology enthusiasts simply assume that a useful device will spread. Where potential customers have access to information and are already served by a private-sector network, spontaneous dissemination may indeed occur. But for the world's poorest farmers, the needed information and private marketing network rarely exist. Dissemination takes special effort.

In 1984, not long after the treadle pump's creation, International Development Enterprises (IDE), a private nonprofit organization headquartered in Denver, Colorado, and with field offices throughout Asia and parts of Africa, became involved in disseminating the new device. IDE's philosophy is to treat poor people as customers rather than as recipients of charity. The organization works to stimulate market demand for products that meet strict criteria of affordability and profitability for the small farmer, and to spawn a self-sustaining network of manufacturers and sellers of the technology.

The treadle pump's spread is largely due to the success of this strategy. IDE worked directly with manufacturers to diversify the production base, acted as a wholesaler to ensure quality control, trained a cadre of local technicians to install the pumps, and set up a network of dealers to sell pumps at a reasonable profit. It spun off a for-profit company called Krishok Bandhu ("Farmer's Friend") that gave the pump a brand name, a logo, and an identity for advertising purposes. Roving musicians, feature films, and open-air dramas spread the word about the technology. In one popular, award-winning film that incorporates traditional Bangladeshi customs, a treadle pump generates enough cash for a dowry that makes a marriage possible. Pump promoters aired the film on large outdoor screens to rural audiences numbering a million a year.[8]

The million-plus treadle pumps now in use in Bangladesh are supported by a network of 73 manufacturers, 830 dealers, and more than 2,500 installers. IDE President Paul Polak estimates that the total market for treadle pumps in the developing world may number 10 million—including 6 million in India, 3 million in Bangladesh, and another 1 million spread among a number of Asian and African countries. The Bangladesh experience shows that spreading the pumps widely takes a good marketing strategy and hard work. In India, where the potential market is twice as large as in Bangladesh, only about 80,000 treadle pumps are in use, in large part because the marketing infrastructure is still in its infancy.[9]

But a great deal hinges on the success of spreading this simple irrigation device. Ten million treadle pumps could add $1 billion a year to small farmers' net income and dramatically improve food security in some of the world's poorest regions. In Bangladesh and elsewhere, the installation of deep tubewells by wealthier farmers poses a major threat to the users of treadle pumps and other small-scale manual irrigation devices. The deeper wells with more powerful pumps can lower the water table below the suction level of the treadle pumps, rendering them useless—just one of the many equity issues that water institutions have failed to deal with adequately. (See Chapter 10.)[10]

Overcoming Barriers in Sub-Saharan Africa

Spreading the benefits of irrigation to small farmers in sub-Saharan Africa is as critical as in South Asia, and even more challenging. Deep pockets of poverty persist in this region. The United Nations estimates that about one third of sub-Saharan Africans do not get enough to eat. Agricultural output in the region grew by less than 2 percent a year over the last three decades, while population rose at close to 3 percent annually. Even fairly conservative projections suggest that cereal imports will need to triple within 25 years—from 10 million tons a year to 30 million. It is highly doubtful that these countries will be able to afford such imports.[11]

Africa's water endowment is rich and varied, but not easily

accessible for conventional irrigation. Only 20 percent of the precipitation falling on land feeds rivers, streams, and underground aquifers, compared with 37 percent globally. The other 80 percent evaporates or is transpired by plants directly back to the atmosphere. This ratio of runoff-to-precipitation is even lower in certain parts of the continent—9 percent in southern Africa, for example, and 6 percent in the Sahel, the band of territory roughly from Senegal and Mauritania eastward to Ethiopia. With so little rainfall turning into runoff, much less of the continent's renewable water can be tapped for irrigation than in Asia or North America. To make matters worse, rainfall is exceptionally variable and unpredictable, and drought is endemic. About a third of sub-Saharan Africans live in drought-prone regions highly susceptible to crop failures. Finally, although groundwater is widespread, it is limited in volume, accounting for only 15 percent of the region's renewable water supply.[12]

Because of these conditions, irrigation has been difficult and expensive to develop in sub-Saharan Africa. New medium and large irrigation projects cost on average $8,300 per hectare, compared with $2,500 per hectare in South Asia. Average costs climb even higher, to $18,300 per hectare, when a variety of associated indirect costs are taken into account, such as the building of roads, electric lines, and other infrastructure in the project area. Shocking stories of $40,000-per-hectare price tags are not uncommon for large government-built irrigation schemes in Africa. The hiring of foreign contractors for design and construction work, the need to import most materials and equipment, heavy taxes on imported goods, and overvalued currencies that inflate all costs in dollar terms contribute to astronomical costs. Because investors face a significant risk of negative rates of return, governments and international donors tend to shy away from large public irrigation schemes.[13]

A new approach is needed if Africa is to enlist the aid of irrigation in boosting food security and incomes. Continent-wide, only 6 percent of Africa's cropland is irrigated, compared with 37 percent of Asia's and 17 percent of the world's. Moreover, half of Africa's 12.3 million hectares of irrigated land is in the northern tier of countries, from Morocco across to Egypt. The whole of sub-Saharan Africa has 6.3 million irrigated hectares, and just three countries—Madagascar, South Africa, and Sudan—account for two thirds of this total. In short, much of Africa has scarcely been touched by irrigation, and most of the poor have missed out entirely.[14]

As in South Asia, lack of access to affordable technologies geared toward poor farmers on small plots has been a major constraint on irrigation's spread. In Malawi, Tanzania, Zambia, and many other sub-Saharan African countries, small-scale farmers account for 80–85 percent of the farming population. Most are subsistence farmers just scraping by, and they face Herculean barriers to irrigation. Equipment often costs 2–10 times more than in Asia, depending on the type and whether it is locally manufactured or imported. Many farmers lack secure tenure to their land, making it risky to invest in improvements. Many smallholders—especially the women, who do much of the farm work in Africa—lack access to credit to purchase supplies and equipment. And poor transportation and marketing facilities often make it difficult to profit from agricultural investments.[15]

Nonetheless, small-scale irrigation schemes have the potential to make a huge contribution in the region. Africa has a long and diverse tradition of indigenous irrigation, but agricultural specialists have tended to underplay this experience because much of it lies outside the definitional bounds of conventional irrigation. Modern engineers restrict "irrigation" to the controlled application of water to crops in a timely manner. In Africa and elsewhere, however, there are many water management techniques that do not fit this definition but that nonetheless provide much more water to crops than natural rainfall would alone. These practices include cultivating wetlands, planting crops in floodplains after floodwaters recede (known as flood-recession farming), and a variety of techniques to capture rainfall and runoff and to channel more moisture into the soil.[16]

Whether such measures count as irrigation depends on who is counting. The Working Group on Small-Scale Irrigation in the United Kingdom, for example, defines small-scale irrigation broadly as "irrigation, usually on small plots, in which small farmers have the major controlling influence, and using a level of technology which the farmers can effectively operate and maintain." Even the U.N. Food and Agriculture Organization (FAO), keeper of the official global irrigation statistics, has begun to publish figures on different categories of water management practices in Africa, thereby acknowledging that controlled irrigation is not the only important game in town. In addition to the 6.3 million hectares of irrigated land in sub-Saharan Africa, FAO reports that there are more than 1 million hectares of cultivated wetlands and valley bottoms and another 1 million hectares of flood-recession cropping. Including these lands increases the total area under some form of water management in sub-Saharan Africa by a third.[17]

Apart from South Africa and the Sudan, most of the countries with sizable irrigated areas use small-scale methods on a majority of their irrigated lands. These range from ancient indigenous techniques using almost no modern technology to shallow tubewells and small diesel pumps. As African irrigation specialist W.M. Adams of the University of Cambridge notes, "Given that almost no resources have been devoted to the support and development of indigenous irrigation, its extent is remarkable."[18]

Both enormous opportunities and major pitfalls exist in building on these traditional experiences and capitalizing on the diversity of Africa's ecological settings. Tampering with indigenous systems to "upgrade" them runs the risk of destroying time-tested social structures and community norms that are integral to their success. On the other hand, the introduc-tion of some new technologies or practices may markedly raise the production levels possible from lands under traditional cultivation, adding to regional food security and incomes.

As in Asia, a variety of low-cost irrigation and water man-agement techniques may be applicable in spreading the bene-fits of small-scale irrigation in Africa. Human-powered pumps, for example, may be well suited to lift water from shal-low groundwater or streams in many eastern and southern African countries. Malawi, Zambia, and Zimbabwe are among those with substantial areas of seasonally waterlogged lands called *dambos*, which collect runoff from higher ground and channel it slowly toward a nearby river. Because *dambos* are fragile wetlands with important ecological benefits, using them to grow crops requires careful management and small-scale technologies that suit *dambo* conditions.

Crops grown on *dambos* typically get 85–90 percent of their moisture directly from the underlying shallow groundwater as capillary action brings soil water directly to the plant. Farmers supplement this natural supply with irrigation water lifted from shallow wells by buckets or handpumps. Depending on local conditions, they sometimes use inexpensive piping to bring surface water to their plots by gravity. Deep wells and modern mechanized pumps would be inappropriate for *dambo* cultivation because they would invariably lower the water table, eliminating the efficient, naturally occurring sub-surface irrigation and damaging the wetlands themselves. But small human-powered devices—such as swing baskets, rope pumps (made locally from a bicycle wheel and rope), and very shallow treadle pumps—can aid the task of irrigating *dambos* and expand the cultivated area.[19]

Dambos cover some 3.8 million hectares in Zambia, 1.3 million hectares in Zimbabwe, and significant areas in several other African countries as well. Most *dambo* plots are less than half a hectare in size and have often been neglected by agricultural officials. Collectively, however, they can amount to a sizable area. The roughly 20,000 hectares of *dambos* under cultivation in Zimbabwe, for instance, equal 13 percent of the nation's official irrigation estimate of 150,000 hectares. Along with careful management practices, the availability of appropriate low-lift irrigation technologies could expand production from *dambos*, improving smallholder food security while sustaining the ecological functions of these seasonal wetlands.[20]

In addition to *dambos*, substantial areas of valley bottomlands are well suited for irrigation with low-cost pumps. In Kenya, for example, a local nongovernmental organization called Approtech reconfigured the Asian treadle pump into a lighter, portable device called a pedal pump for irrigating small valley-bottom plots. Nicknamed "the moneymaker," the pedal pump sells for about $70 and is actively marketed in Kenyan towns and villages. In a two-month period after the pedal pump's introduction in 1996, Approtech sold more than 300 pumps, and demand quickly began to outpace supplies.[21]

In northern Nigeria, farmers are tapping shallow groundwater with low-cost wells and motorized pumps for dry-season irrigation of seasonally flooded river valley areas called *fadama*. These pump systems typically irrigate 1–2 hectares and cost $350–700 per hectare—considerably more than the manual pumps, but far less than most large public-sector schemes. They were introduced in the 1980s as part of a larger World Bank agricultural loan package, and turned out to be one of the few successful project components.[22]

Yields of rice and maize—basic staples in the region—rose markedly, as did those of onions, peppers, tomatoes, and other marketable crops. As a result, both incomes and food security improved for families farming in the *fadama*. Some 50,000 of these low-cost irrigation systems are now operating in Nigeria, and a strong support-service industry has evolved to maintain them. As with the treadle pump, this service network not only creates jobs and strengthens the local economy, it helps ensure that the irrigation systems will spread and be maintained long after their initial introduction. Evidence of localized over-pumping, however, raises questions about the long-term sustainability of these farming practices and underscores the need for regulations that limit groundwater use to the level of replenishment. (See Chapter 10.)[23]

Small-scale pump and tubewell systems have also met with considerable success in Chad and Niger, in the heart of Africa's Sahel. Soon after farmers in Chad heard about the low-cost pumps available in Nigeria, they began making deals with local merchants traveling there to bring back pumps and spare parts. While manual water lifting may allow for one cash crop in the dry season, the mechanized pumps allow for at least two, and sometimes three crops. Farmers able to afford the shift from manual lifting to the small, mechanical pumps have seen their cash income more than double and their person-days of labor drop by 60 percent.[24]

If small irrigation pumps combined with tubewells to tap shallow groundwater are so successful, why have they not spread more rapidly in Africa? The answer, according to Ellen Brown and Robert Nooter, is that these technologies are simply unknown to many Sahelian farmers. In Chad, for example, the small pump-and-tubewell systems began to spread only after a few farmers had seen them used in Nigeria and after an inexpensive method was devised for drilling boreholes. Once that knowledge was gained, however, use of the technology spread rapidly. "Extension services need to publicize and demonstrate technologies at the farm level," Brown and Nooter write, "not just on demonstration farms that few farmers ever see." Other important barriers that these and other researchers have found include difficulties in obtaining credit and the absence of secure land rights—especially for women.[25]

Finally, modern enhancements to traditional practices of flood-recession agriculture show promise in parts of Africa. As the phrase implies, flood-recession farming involves planting crops after a river's seasonal flood recedes. The moisture stored in the floodplain soils supports the plants throughout the growing season. Farmers practicing this method typically trade higher yields for greater drought resistance and early maturation of crops, because if the soil moisture runs out before harvest time, they can lose their entire crop.

Flood-recession agriculture is an age-old practice in Africa, extending back at least 5,000 years in the case of Egypt's Nile basin. The presence of early-maturing and drought-resistant varieties of maize, millet, and sorghum suggests that flood-recession cultivation was historically widespread on the continent. The eleventh-century historian al-Bakri described the practice in the Senegal River valley: "The inhabitants sow their crops yearly, the first time in the moist earth during the season of the...flood."[26]

Today, in good years, between 300,000 and 500,000 hectares are used for flood-recession cultivation in the Senegal, Niger, and Lake Chad basins. Judged by grain yields alone, this cultivation method appears considerably less productive than modern intensive irrigated agriculture. Typical cereal yields range between 400 and 800 kilograms per hectare, compared with 2,000–4,000 kilograms per hectare on irrigated lands with good water control. But this conventional agricultural measure fails to capture other highly productive elements of a flood-recession system—elements that together contribute to much higher water productivity than grain yields alone suggest.[27]

Most farmers, for example, plant legumes and vegetables along with their grains. Nourishing grasses and shrubs that grow on the floodplain provide a source of pasture for livestock herders during the dry season. Anthropologist Thayer Scudder points out that many farmers welcome the migrating pastoralists because their herds leave behind manure that fertilizes the floodplain soils. The annual flood also enhances the productivity of riverine fisheries by providing habitat for spawning and protection of young fish. And active floodplains are typically easier for local people to fish than main river channels, increasing access to high-quality protein in areas where nutritional deficiencies might otherwise occur. The seasonal flood sustains vital wetland habitat for wildlife populations and recharges groundwater aquifers that rural people depend on for drinking water and small-scale irrigation.[28]

Trained to aim their project plans at raising crop yields, most water and development planners overlook the multiple production benefits of these traditional flood-based systems. A case in point is the Manantali Dam, completed in 1986 on Mali's Bafing River, a main tributary of the Senegal River, which is shared by Mali, Mauritania, and Senegal. With international aid, engineers began constructing the dam in the 1970s after a devastating drought in the African Sahel. The dam's main purpose was to expand irrigation, generate hydropower, and extend barge transportation. By eliminating the seasonal floods, however, operating the dam would destroy a highly productive flood-based system that downstream river valley inhabitants depended on for their livelihoods.[29]

In a creative example of ecologically oriented engineering design, an international, interdisciplinary research team spearheaded by the Institute for Development Anthropology (IDA) at the State University of New York at Binghamton demonstrated that the primary goals of the dam could be met without destroying the flood-dependent downstream production systems. First, they showed that the dam would store enough water for hydropower generation as well as for the release of a controlled flood to sustain the traditional riverine systems. Second, they showed that when all the benefits were taken into account, the traditional flood-based system was actually more productive than the conventional irrigation option. The researchers found that each hectare of floodplain could support 5–10 times more livestock than a hectare of nearby rangeland, and that the Senegal valley typically yields some 30,000 tons of fish a year. According to IDA director Michael Horowitz, "the net returns per unit land from the total array of traditional production—flood-recession farming, herding, and fishing—actually exceed those from irrigation, without taking into account the latter's huge start-up and recurrent operating and maintenance costs."[30]

Convinced by the team's findings, the Senegalese government endorsed the strategy of a controlled "artificial" flood. But the Senegal Valley Development Authority, which includes all three countries in the basin and is responsible for operating Manantali Dam, has continued to insist that the dam be operated to completely eliminate the flood. Both Mauritania's policies and upstream Mali's primary interest in hydroelectric revenues are major barriers to the acceptance of the controlled-flood option. Meanwhile, a union of Senegal valley farmers has begun to speak out, calling for the reestablishment of the flood to safeguard their crop production, livestock herding, and fishing.[31]

Whatever the outcome, the Manantali case establishes new possibilities for river-based agriculture, renews and modernizes the concept of ecologically based irrigation initially practiced by the ancient Egyptians, and offers a model of water development that benefits the poorest farmers. Researchers are already exploring the relevance of this approach to river basins in northeastern Nigeria, the Tana basin in Kenya, and the Mekong in Southeast Asia.

Small, Affordable, and Efficient

For many small farmers, getting access to water for irrigation is less of a problem than stretching the supply available to them. Despite worsening water scarcity in many parts of the world, modern pressurized irrigation systems that have great potential to reduce water use—including sprinklers, drip, and other microirrigation methods—have spread relatively slowly in most parts of the world. With the advent of new low-cost drip irrigation and low-cost sprinklers, however, many small farmers can irrigate more land with their limited water and get better yields, often doubling their harvests. (See also Chapter 8.)

In recent years, for example, a spectrum of microirrigation systems has slowly begun to spread, each one incorporating the fundamental water-saving drip principles into a system affordable and appropriate to a particular farming area. The simplest and least expensive drip system is the bucket kit, which is ideal for irrigating the home gardens that provide food security for many rural smallholders. The kits have a simple 20-liter bucket installed on a vertical pole at shoulder height, with a plastic lateral line stretching out from the bucket. Extending from the lateral line are 26 microtubes—narrow flexible tubes that deliver water to the soil in small quantities. By strategically placing them midway between two rows of crops, each microtube can irrigate four plants. Systems costing as little as $5–10 can water 100 plants, about the number in a typical home garden for a six-person household. At shoulder height, the bucket has enough head to deliver water to the entire garden, and, if filled twice a day, can satisfy all the crops' water needs. If the garden had previously been watered by hand, the bucket kit should save labor as well as water.[32]

One step up from the bucket kit is the drum kit, consisting of a 200-liter drum and a larger network of laterals and microtubes. Costing about $25, these kits can typically water a 100-square-meter plot, or about 400 plants. Further up the drip ladder are shiftable systems costing $100 for a fifth of a hectare (a half-acre), and useful in areas of hill agriculture. Stationary microtube systems are at the upper end of the low-cost spectrum. Farmers find these systems useful where labor costs are relatively high or where shifting the system is difficult. As with the simple bucket kits, strategic placement of the microtubes allows each lateral line, an expensive component, to irrigate four rows. Such cost-saving features keep the price around

$125 per fifth of a hectare, about a quarter as much as a conventional drip system.[33]

Just as exciting as the development of low-cost drip irrigation is the emergence of low-cost sprinklers. Sprinklers water less than 10 percent of the world's irrigated area even though they can be used on a variety of widely planted crops, including wheat, oilseeds, pulses (legumes such as beans and lentils), and many vegetables. As with drip irrigation, developing and marketing low-cost sprinklers for small-scale farmers is a key to expanding the global area.[34]

Various low-cost sprinklers have begun to enter the marketplace. One overhead system includes a sprinkler head on a tripod, which a farmer could move eight times in order to irrigate about half a hectare. These sprinklers sell for $60, not including the water delivery system. A second type includes lateral lines (similar to a drip system) fitted with a series of microtubes that each have a little rotating sprinkler head attached to them. These microsprinklers are too close to the ground to irrigate wheat, but are useful for vegetables and other closely spaced crops. The network of laterals and tubing can be moved from plot to plot, allowing one system to irrigate more than one parcel of land. By August 1998, small farmers in India had purchased some 250 low-cost sprinkler systems—a number that seems likely to increase rapidly as more of them learn about the technology.[35]

Harvesting the Rain

There is yet another category of small farmer who lacks access to irrigation—the millions in arid and semiarid areas where there is no perennial water supply. Without a water source to tap during the dry season, these farmers cannot benefit from irrigation. Unreliable and insufficient rainfall is the only source of water for their crops, and many suffer crop failures in one out of three years. More stable harvests are critical to the food security and livelihoods of these farmers, who rank among the world's poorest. Where shallow groundwater, perennial streams, or other water sources are not accessible or sustainable, a set of practices known as water harvesting may offer the best hope of getting more water to crops.

On most dryland farms, whether in India, China, sub-Saharan Africa, or elsewhere, only 15–30 percent of rainfall actually adds moisture to the root zone and thus aids crop production. Typically, 20–25 percent runs off the field, another 20–25 percent drains deeper through the soil, and 30–35 percent evaporates from the soil. Channeling a greater share of rainfall into the soil, and conserving moisture in the root zone where crops can use it, can help safeguard harvests from drought and raise crop yields overall. If irrigation is defined as any method that supplies more water to crops than natural rainfall would alone, these practices constitute a form of irrigated agriculture.[36]

In India, disappointing results from many large-scale irrigation projects along with mounting opposition to the construction of large dams is rekindling interest in a rich variety of traditional, precolonial irrigation practices. These various water harvesting methods developed over the centuries to enable villagers to cope with large seasonal differences in rainfall. India gets 80 percent of its precipitation in three to four months, making rain-fed crop production during the other eight to nine months highly risky. The diversity of water harvesting methods that evolved matches the diversity of India's ecological conditions and cultural traditions, and many have withstood the test of considerable stretches of time.[37]

In the upper part of the Narmada River valley in Madhya Pradesh, a little-known traditional water-harvesting method called the *haveli* system still functions, although it is now threatened by the controversial series of large dams slated for construction on the Narmada River. (See Chapter 4.) Rainwater from the wet season is held in fields by earthen embankments constructed on all four sides. Locally called *bundhans*, these embanked fields vary in size from 2 to 12 hectares. If the monsoon rains are good, they will typically fill with water by August. In early October, farmers drain the water off, and then plant their crops. The moisture remaining in the soil is sufficient to sustain the crops until harvest time. In this way, farmers can produce a crop during a winter season that would otherwise be far too dry.[38]

Covering an area of about 140,000 hectares in various districts of Madhya Pradesh, the *haveli* system, like most water harvesting methods, is tailored to the environmental conditions of the region. It works because the heavy, black, clayey soils common to this area can hold large quantities of water. Although the system precludes growing paddy and other monsoon crops, because the fields are used for water storage during this time, it is well suited for a variety of other crops, including wheat, lentil, and linseed. As with many traditional water harvesting systems, a critical element of success is a social structure and set of community norms and practices to which farmers adhere. In the *haveli* system, there is a mutual understanding among the farmers as to when to drain their *bundhans*, since the runoff flows by gravity from one field to the next until it joins the natural drainage downstream.[39]

Today, less than half of the *haveli* system remains intact. Many farmers have shifted to soybeans, a monsoon crop that precludes using the fields to store water. Ironically, some form of the *haveli* system may be essential to the sustainability of the modern methods that are replacing it. The *haveli* areas have become one of India's largest sprinkler-irrigated tracts, but without the groundwater recharge resulting from inundating the fields for several months a year, water tables are declining in many areas. By keeping enough area under the traditional system to replenish groundwater supplies, and using efficient sprinklers—including micro-varieties—to irrigate during the dry season, the traditional and modern approaches together might work better than either alone.[40]

In many drier parts of India, traditional agriculture depended on tanks—small reservoirs of varying sizes that store rainfall or runoff from the wet season for use during the dry season. Some 2 million tanks dot India's landscape, and they collectively irrigate about 3.5 million hectares. In the southernmost state of Tamil Nadu, ancient tanks known locally as *eris* water a third of the irrigated land. Besides storing runoff for use during the dry season, the *eris* help control flooding, reduce soil erosion, and promote groundwater recharge. In larger watersheds, they may be arranged in a cascade fashion, with the overflow from one feeding the next one downstream.[41]

Until the British arrived in 1600, the *eris* were maintained mostly by local villages themselves, aided from time to time by infusions of resources from a king, local warlords, or chiefs. Each village allocated about 5 percent of its gross production for maintenance of the tanks and other irrigation structures, and collectively they supported those responsible for doing this work. As a result, the *eris* were desilted regularly, and the sluices and irrigation channels were kept in good working order.[42]

With the introduction of early British rule, these social structures began to collapse. The large expropriation of village resources by the state left little to support the water harvesting systems. In the mid-nineteenth century, the *eris* were brought under the centralized control of the state public works department, which allocated little money for their upkeep. Realizing that cooperation from the villages was essential to the functioning of the tank systems, the state attempted to force that cooperation by passing the Madras Compulsory Labour Act of 1858, but to no avail.[43]

After independence from the British, little changed, since centralized control persisted under the new Indian political system. In both cases, that control stopped at the tank itself. The villages retained control over how the water in the tanks was allocated and used. Informal farmer organizations existed to make these decisions, and these organizations have persisted to the present. Some are now calling for the *eris* to be returned to village control.[44]

At a minimum, the revival of India's traditional tank systems depends on a careful evaluation of whether the conditions that sustained them in the past are practical and can be recreated today—including reestablishing the villagers' active participation in tank management and upkeep. Many tanks now lie in disrepair in degraded watersheds. Bringing them back into service would require substantial efforts to control soil erosion by better managing the surrounding land. In the state of Karnataka, for example, where more than 36,000 tanks have been constructed over many centuries, silt has destroyed 30 percent of the total tank capacity.[45]

Despite the difficulties, efforts to revitalize the use of tanks is an ecologically sound response to stabilizing and improving crop production in dryland regions where many millions remain poor and underfed. Studies in India's drylands have shown that even one irrigation on otherwise rain-dependent lands can increase yields of sorghum and upland rice by 40–85 percent and can double yields of wheat. Tanks help recharge the underlying groundwater, making pumping systems more sustainable. As with the *haveli* system in Madhya Pradesh, promising opportunities may exist to marry traditional tank systems with modern low-cost drip and micro-sprinkler systems, thereby stretching the limited water supply and boosting water productivity.[46]

These descriptions only hint at the rich diversity of water harvesting systems in India, not to mention in other regions. In their book *Dying Wisdom*, Anil Agarwal and Sunita Narain of the New Delhi–based Centre for Science and the Environment have pulled together a comprehensive compendium of traditional Indian water harvesting methods. Noting that these systems can be built in months instead of years, require relatively small sums of money, can be locally controlled, and can be used jointly with modern tubewells, they see a new emphasis on these methods as at least a partial alternative to expensive and increasingly controversial large-scale projects.[47]

Another set of techniques, called "runoff agriculture," offers a variation on water harvesting. Runoff agriculture was practiced by ancient agriculturalists as many as 3,000 years ago, and can allow crops to thrive in very dry regions where production would otherwise be impossible. The practice involves channeling rainwater from the surrounding land area to a field of crops. In very dry locations, this collection area—called a catchment—may need to be many times larger than the farm plot in order for the crops to get enough moisture to reach maturity.

In Israel, for example, researchers are reviving techniques practiced by the Nabateans, caravan traders who occupied the Negev Desert some 1,600–2,000 years ago. Just below the hilltop remains of the ancient Nabatean city of Avdat, researchers at the nearby Jacob Blaustein Institute for Desert Research have a 1-hectare farm watered by a 250-hectare catchment. The runoff from a few heavy rains during the wet season is channeled to the field, which is enclosed by small stone walls. The plot can store up to 250 millimeters of water, enough to sustain the crops through the dry season. The researchers are experimenting with various ways of improving upon ancient Nabatean practices. They have found, for example, that covering the soil between crop rows with polyethylene cuts down evaporation and, by heating up the root zone, causes plants to use water more efficiently. Yields of sorghum have doubled with this method, from 1.5 tons per hectare to 3 tons.[48]

In many watersheds, constructing a small dam across a gully can trap large amounts of runoff, which can either be channeled directly to a field, stored for later use in a tank or small reservoir, or allowed to percolate through the soil to recharge the underlying groundwater. Often called check dams, temporary structures can be made of loose rocks, earth, woven wire, or other locally available materials. They typically last two to five years and cost $200–400, depending on the materials used, the size of the gully, and the dam's height. More-permanent structures made of stone, brick, or cement typically cost $1,000–3,000.[49]

About 40 kilometers south of Indore in the central Indian state of Madhya Pradesh, I visited a promising effort to boost crop production in a 500-hectare watershed that is home to some 80 families. Along with a check dam to trap runoff, it includes a percolation pond—a constructed basin where the trapped water slowly seeps through the soil to recharge the underlying groundwater. A look across the landscape showed the early fruits of success: fields of wheat, vegetables, and fruit trees on what was previously marginal grazing land. With the water table now high enough for the villagers to tap groundwater with small motorized pumps, they have begun to irrigate portions of the watershed.[50]

Once again, there is great potential to increase the productivity of these water harvesting schemes through the use of low-cost sprinkler and drip systems to deliver the water to crops more efficiently. At the site near Indore, I saw rainwater successfully captured through labor-intensive watershed efforts gushing out of a pipe to flood-irrigate wheat fields. The use of affordable and efficient irrigation systems in these watershed areas would help stretch the water supply so painstakingly captured—lifting water productivity, local food production, and incomes.

Most water harvesting schemes require communal work and a high degree of cooperation and organization if they are to function properly. These requirements can make revitalizing tanks, building check dams, constructing bunds and terraces, and other watershed projects difficult and relatively expensive to carry out, compared with small-scale schemes based on manual or small mechanized pumps that farmers can purchase and operate individually. This is especially the case where traditional community customs are no longer in effect, so that mechanisms for farmer participation and management must be rebuilt. Moreover, watershed projects often raise questions about who benefits and who loses, just as large dams do. In some cases, for example, farmers with land may benefit while pastoralists and women lose access to grazing lands and fodder for their animals.

Werner Hunziker, head of natural resources management in the New Delhi office of the Swiss Agency for Development and Cooperation, which funds many watershed projects in India, notes that there is still a struggle to find the proper institutional arrangements to do effective and efficient watershed development. There is no assurance, he says, that it will be possible to take the projects done so far and scale them up. Yet with average costs in India of about 10,000 rupees ($235) per hectare, watershed schemes fall well within the range of cost-effective small-scale options. In dry regions with no year-round water supply, these schemes may be the only hope for improving food security and the lives of poor farm families.[51]

10

THE PLAYERS AND
THE RULES

===

*We have learned that implementing reform on a
timely basis is far more complicated than pouring concrete.*

George Miller
U.S. Representative

Say the word "irrigation," and more likely than not what
springs to mind are images of dams, reservoirs, canals, rotating
sprinklers. They are images of irrigation's hardware—the physical infrastructure that delivers water to farms. But just as a
computer is only as good as the software that runs it, so the
benefits of an irrigation system depend on the rules that determine how that system works. For most of the twentieth century, water managers and engineers have focused on building
irrigation's hardware. The development of good software, on
the other hand, has lagged badly behind. In the computer
world, it would be as if IBM had never met Microsoft.

If we are to meet the challenge of feeding 8–9 billion people in a world of increasing water scarcity, we need to upgrade
irrigation's social capital along with its physical capital. From
the wealthiest farms in California to the poorest ones in
Bangladesh, the game of irrigation involves many players, each
of whom behaves according to a set of rules and incentives.

These players include farmers, irrigation districts, water user organizations, state or provincial water agencies, national ministries, development banks, aid agencies, private voluntary organizations, engineering firms, politicians, and taxpayers.

The rules, especially for large schemes, often look something like this: A politician seizes the potential for bolstering political support by proposing an irrigation project in a strategic location. Engineering firms lobby the decisionmakers in order to raise their chances of winning the project's construction contract. The politician, in collusion with colleagues who also want to please their constituents with pork-barrel projects, sees to it that the nation's taxpayers foot most of the bill. The farmers themselves pay only a small fraction of the project's cost; in return, they support the politician in the next election.

With prices kept artificially low, farmers have little incentive to use water efficiently. The irrigation system never becomes financially self-sustaining because the meager fees collected from the farmers do not cover the system's operation and maintenance costs, much less its capital costs. National or state irrigation agencies keep the projects running with taxpayer funds. If budgets become tight and maintenance work is neglected, the systems fall into disrepair. Gradually, agricultural output and benefits begin to decline. Either the government saves the project by allocating more taxpayer money for expensive rehabilitation work, or the system deteriorates until farmers abandon it. Alternatively, international donors come to the rescue with funds provided by taxpayers in wealthier countries.

Variations on this simplified story explain the creation and operation of most large irrigation schemes in the world today. Government motives for building irrigation projects are sometimes more laudable than those of the vote-seeking politician, but the end result is more or less the same. The rules of the irrigation game are stacked against efficiency, fairness, and sustainability. Artificially cheap water encourages wasteful use by farmers, which in turn contributes to water depletion and land degradation. The politicization of irrigation decisions favors wealthier farmers and politically connected engineering firms. The failure to collect adequate fees prevents the irrigation scheme from ever becoming financially sustainable.

The Landscape of Incentives

Signs of irrigation's flawed software design are easy to find. Large subsidies for irrigation are nearly universal, for example. Farmers receiving water from government-built projects rarely pay more than 20 percent of the water's real cost, and often much less. The Indian government had recovered less than 10 percent of the total recurring costs for major- and medium-sized irrigation projects built by the mid-1980s. Farmers in Tunisia, a country severely short of water, pay 5¢ per cubic meter for irrigation water, one seventh what it costs to supply it. And irrigators in Jordan, one of the most water-strapped nations in the world, pay less than 3¢ per cubic meter—a small fraction of the water's full cost. Environmental consultant Norman Myers estimates that irrigation subsidies worldwide total at least $33 billion a year. If the full costs of environmental damage, human resettlement from dam sites, and increased water-borne disease from irrigation projects were factored in, the total subsidy would tally far higher.[1]

In the United States, where the Bureau of Reclamation supplies water to about a quarter of all irrigated land in the West, the subsidy situation is no better. The 1902 Reclamation Act, which aimed to settle the western frontier by helping family farmers obtain water and power, set in motion a process of rent-seeking—the search for profits above and beyond what a competitive market would provide—that persists to this day. Rent-seeking takes various forms in various places, from direct bribes and corruption to collusion among farmers and influential politicians to ensure that irrigation costs continue to be paid by taxpayers. In much of the western United States, these practices resulted in a deeply entrenched and pervasive pattern of subsidies. At various points in time, the Bureau of Reclamation decided not to charge their irrigation clients any interest on water project construction costs, to extend the deadline for repayment of costs, and to limit repayment obligations to an ill-defined notion of farmers' "ability to pay."[2]

Over time, subsidies ballooned. Wealthy farmers often dipped deepest into the public trough. In a study of rent-seeking in irrigation, resource economist Robert Repetto found that "farmers who have successfully captured, defended, and increased the rents implicit in federal irrigation policy are among America's richest." Repetto placed the average subsidy to beneficiaries of federal irrigation largesse at 83 percent of full project costs—a total of more than $1 billion a year. The U.S. Congressional Budget Office calculates that from the inception of the federal irrigation program in 1902 through the mid-1980s, irrigation subsidies totaled between $33 billion and $70 billion.[3]

The financial terms of two recent U.S. government projects provide a striking lesson of how outlandish irrigation welfare can become. The Central Arizona Project (CAP), completed around 1993, diverts water from the Colorado River and delivers it by canal to portions of Arizona. Officials set the contract price for deliveries of CAP water to an irrigation district in the central part of the state at $2 per acre-foot (less than two tenths of a cent per cubic meter), although the full cost of that water was $209 per acre-foot. In other words, the government charged these desert irrigators just 1 percent of the actual cost of the water.[4]

Officials set strikingly favorable terms for the Central Utah Project (CUP) as well. The CUP consists of a series of dams, tunnels, and aqueducts to transfer water from Colorado River tributaries into Utah's Great Basin. Delivering CUP water to Utah farmers costs an estimated $400 per acre-foot. The estimated value of that water to Utah farmers—the maximum amount they would be willing to pay for it under market conditions—is only $30 per acre-foot. The farmers will actually be required to pay only $8 per acre-foot—2 percent of the water's real cost and nearly three quarters less than it is worth to them.[5]

Along with massive subsidies, another major disincentive to efficiency is the absence of accountability for how an irrigation system actually performs—a problem that is especially serious in developing countries. Irrigation fees often go into the national treasury rather than into a fund earmarked for operating and maintaining the system from which the fees were collected. As a result, the fees farmers pay often have no bearing on the upkeep and performance of their irrigation networks. Understandably, many irrigators simply choose not to pay at all, and become "free riders." For their part, irrigation agency staff have little incentive to collect fees and improve operations because their job security is rarely tied to how well the systems work. With incentives almost completely divorced from performance, many public systems operate poorly.

One of the most costly and damaging consequences of these perverse incentives is the notorious tail-ender problem. Farmers at the head of a canal system take as much water as they want, leaving farmers at the tail-end with little or nothing. While soils are getting waterlogged at the head, crops are withering from dryness in the tail. University of Sussex researcher Robert Chambers cites cases in South Asia where farmers in the head-reaches got five irrigations while tail-enders barely got one.[6]

With water so scarce and unpredictable, farmers in the lower stretches of canal networks plant less-risky crops and apply less fertilizer. As a result, yields typically decline from head to tail, and so do farmers' incomes. One study in India found that irrigators near the head of the canal system earned six times as much as those in the tail. Moreover, the size of the problem is staggering in some food-producing regions. Chambers has estimated that 25–40 percent of the area declared irrigated in India suffers from tail-end deprivation. The lost productivity, reduced income, increased inequity, and resource degradation caused by the skewed distribution of water in canal systems make the tail-ender issue one of the most serious unsolved problems in irrigation.[7]

The whole panoply of perverse incentives driving irrigation operations has mushroomed with the infusion of funds made available by the World Bank, regional development banks, bilateral aid agencies, and other international donors. Between 1950 and 1993, the World Bank alone lent $31 billion for irrigation projects. Most of the money was channeled into the construction of large, capital-intensive schemes. Relatively little went into creating the social capital—laws, policies, and institutions—that was needed for the large infrastructure to function properly. As a result, irrigation's overall performance has been decidedly mixed: It contributed to impressive increases in crop production, but also to an expensive set of failures and worsening inequities that still plague irrigated agriculture today.[8]

The Mahaweli scheme in Sri Lanka, for example, involved construction of four large dams on the Mahaweli River and aimed to boost hydropower production and supply irrigation water to some 364,000 hectares. Aided by credits and loans from a number of donors, including the World Bank and the U.S. Agency for International Development, the Sri Lankan government poured massive funds into the scheme. At its peak, the Mahaweli project was absorbing 44 percent of the country's public investment expenditures.[9]

Not long after the scheme's completion in the early 1980s, numerous management problems turned up. One study found that only half the farmers in the project area were getting their irrigation water from authorized canal outlets. The other half were illegally diverting canal water or relying on drainage from other farmers' fields. Instead of following an orderly rotation system, irrigators blocked and breached ditches and outlets in chaotic attempts to get more than their fair share of water. One farmer was found to be siphoning off the entire flow of an irrigation ditch, leaving others high and dry. Robert Chambers points out that the die was cast from the earliest planning stages for Mahaweli, with a key document devoting scarcely a page to how the irrigation system would be managed. Nor did Sri Lankan professionals step in to fill the gap. One engineer expressed his idea of management responsibilities as follows: "'You open the sluice, and you go to sleep.'"[10]

Given this neglect of fundamental management issues, it is hardly surprising that many project planners failed to even consider less obvious but vitally important issues—such as how large irrigation schemes would affect women and households. Because income from irrigated crop production would often go to men, who in many societies spend a smaller share of their incomes on their families than women do, a shift in land use from rain-fed to irrigated crop production could have the twisted effect of raising crop yields but reducing household food security.

This outcome occurred in portions of the Mahaweli scheme. Women there had traditionally cultivated millet and other rain-fed crops for household consumption. Upon completion of the irrigation scheme, much of this land shifted into irrigated rice production for cash income and came under the control of the men. Without access to their husbands' incomes and having lost their rain-fed lands, women could no longer provide sufficient food for their families. In one project area, chronic malnutrition rates reportedly rose to 38 percent compared with a national average of 7 percent.[11]

Irrigation planning and construction also has been a very top-down affair. Governments designed and built many large projects without ever soliciting the ideas and concerns of farmers. A recent World Bank assessment of its role in irrigation notes that evaluation records of Bank-funded projects contain little information on farmer participation and seldom even mention gender issues. Not surprisingly, irrigators have often felt no ownership in the systems built to serve them. Many choose not to abide by rules and institutional arrangements that were crafted without taking their specific needs into account. Even worse, operational rules handed down by irrigation bureaucrats sometimes destroy traditional norms and practices that had functioned well for centuries.[12]

The failure of governments and international donors to design institutions—rules of the game—that would ensure the efficiency, equity, and sustainability of irrigation schemes has left a large backlog of unfinished business in irrigated agriculture. "Over the next several decades," writes political theorist Elinor Ostrom, "the most important consideration in irrigation development will be that of *institutional* design—the process of developing a set of rules that participants in a process understand, agree upon, and are willing to follow.... Social capital is not automatically or spontaneously produced. It must be crafted."[13]

Water Pricing, Water Marketing

The most important rules of the irrigation game are those establishing how water is allocated—who gets how much and when—and how irrigation systems are managed—who has responsibility for operating and maintaining them. As water for agriculture becomes scarce, as new irrigation projects become prohibitively expensive, and as government after government struggles with fiscal crises that necessitate budget cutbacks, the search is on for better ways to do both.

Economists claim that pricing irrigation water closer to its full cost is theoretically the neatest way to promote efficient use and allocation of the resource. Actually raising water prices, however, can be a political high-wire act. Not only does pricing reform require bucking entrenched interests, it runs the risk of pricing poor farmers out of the irrigation game entirely.

There are also practical barriers to charging farmers the full cost of the water they use. Most irrigation systems do not have equipment that measures the volume of water delivered to small, individual farms. More often than not, installing such equipment would cost more than it would yield in benefits. The way many irrigation systems are constructed, managers cannot even cut off water deliveries to a nonpaying farmer. In addition, if water fees go to the national treasury rather than to a fund for maintaining that particular system, higher prices will not result in better service and so farmers will not support them. Many studies have shown that farmers are able and willing to pay more for their water, but only if deliveries become more reliable and service overall improves.[14]

Where it is not practical to charge farmers the real cost of the water they actually use, it is nonetheless possible to design irrigation operations so that water gets used more efficiently and allocated more fairly. For many decades, the 14 million hectares of irrigated land in the Punjab of northwest India and Pakistan—the world's largest contiguous irrigation network—has operated according to a system of fixed turns called *warabandi*. The volume of water available in the Punjab is sufficient to fully irrigate only about a third of the total system area, so a method of optimizing water use under those scarcity conditions is key to maximizing water productivity. Originally applied in colonial India in the late nineteenth century, *warabandi* is designed to ration scarce water in an equitable manner and to spread the risks of water scarcity fairly among all farmers in a canal network.[15]

Warabandi operations involve a three-step process. First, irrigation system managers post a schedule of canal openings and closings that rotates every eight days throughout the irrigation season. In this way, each canal has equal probability of receiving water as flows in the main channel fluctuate. Second, the volume of water delivered to each subgroup of farmers is proportional to the crop area served so that each group gets a fair share of the available water. Third, every farmer within each subgroup is scheduled to start and stop irrigating at a unique time of the week. Officials determine the length of an irrigation turn by the size of the farmer's landholding, again ensuring that an equal volume of water goes to all crop areas.[16]

Warabandi systems are less than ideal. They force farmers to adjust their cropping patterns to the water supply, when ideally it should be the other way around. In theory, systems able to deliver water on demand are preferable because they give farmers more flexibility in their crop choices and farming practices. Farmers with their own groundwater pumps enjoy this kind of reliable and flexible supply, for instance, and they are often willing to pay quite a bit more for it. In practice, however, it is difficult to achieve on-demand service for many small farms in a large canal network.[17]

The *warabandi* method, on the other hand, seems to work. It services a contiguous canal network encompassing fully 6 percent of the world's irrigated land. Together with groundwater wells, the system has helped turn this corner of South Asia into a highly productive breadbasket. Where the conditions exist for this kind of fixed-turn system to function well—including clear landholdings, good canals and field channels, physical water scarcity, and an ability to maintain a constant flow at the main outlet—it offers a method of allocation that promotes relative efficiency, equity, and stability. It also inspires a key lesson for crafting irrigation rules: look for what works rather than what is theoretically correct, and do not let the perfect become the enemy of the good.[18]

Where it is possible to set fees according to how much water farmers use, a broad spectrum of options exists between full-cost pricing, which may be politically unpalatable in the short term, and a marginal cost of near zero to the farmer, which is a clear invitation to waste water. In 1992, the U.S. Congress passed legislation overhauling operation of the Central Valley Project, a large federal irrigation scheme in California. Among other things, the act begins to cut away some of the heavy subsidies these irrigators have enjoyed for so long. The new water pricing structure calls for prices to remain the same for 80 percent of an irrigation district's water allotment, to increase to halfway between current rates and the full water cost for the next 10 percent, and to rise all the way to the full cost for the last 10 percent. Although this method of charging is a far cry from full-cost pricing, it does encourage farmers to conserve water and use it more efficiently so as to avoid the higher unit prices.[19]

Many regions are turning to water markets as a better way to allocate scarce water among competing users. When farmers can buy and sell water freely, they have an incentive to choose more carefully how they use their allotted supply. If they irrigate their crops more efficiently, for example, they can increase their incomes by selling the water they save. If they switch to a less thirsty crop, they can sell the water they no longer need. Or they can decide to stop irrigating, either temporarily or permanently, and sell their irrigation water to someone else.

Formal water markets only work where farmers have legally enforced rights to their water, and where those rights can be traded. In the western United States, for instance, where a "first-in-time, first-in-right" system called prior appropriation prevails, most irrigators have water rights that are as firm as property rights. Water entitlements come with a date that determines the pecking order of rights-holders in a river basin. A farmer whose rights date to 1888 will get water before a neighboring farmer with a date of 1901. Senior rights are very valuable in times of drought, because they will be honored in full before holders of more junior rights get even a drop.

Until fairly recently in the United States, a variety of federal and state laws and regulations inhibited farmers from selling their water. Under the "use it or lose it" rule, for example, irrigators could lose water rights that they did not put to some use considered "beneficial." As legislatures and courts gradually sweep away these restrictions, markets are opening up. The state of Colorado, for instance, has one of the best institutional support networks for water markets of any state, and trading has flourished there—both within irrigation districts and between cities and farms. Farmers within the Northern Colorado Water Conservancy District, which receives water from the federal Colorado–Big Thompson project, can freely transfer their annual entitlements to others within the district. In recent years, about 30 percent of the district's annual water entitlements has moved through the rental market.[20]

In California and in the water-tight Colorado River basin, water marketing is picking up as well. The same congressional act that restructured pricing in California's federal Central Valley Project also explicitly encouraged voluntary water trading. In March 1996, the Westlands Water District, which is supplied by this federal project, began operating an electronic bulletin board for water trading. This mechanism enables farmers within the district to buy and sell their water entitlements from a home computer hooked into the trading system.[21]

As the thirstiest of the seven states in the Colorado River basin, California is looking to water markets to ease its impending problem of shortages. As described in Chapter 6, water deals between the Imperial Irrigation District (IID) and both the Metropolitan Water District in Los Angeles and the city of San Diego will shift water out of agriculture to these burgeoning cities. IID's water use accounts for more than two thirds of California's legal entitlement to Colorado River water—an untenable anachronism in a state that has seen its population swell past 30 million, but legally difficult to do anything about.

Now that U.S. Interior Secretary Bruce Babbitt, water master of the Colorado River, has called on California to stop living off other states' unused river allocations, IID's vast pools of irrigation water are a precious commodity. In late 1997, Babbitt announced that the federal government would allow trading of Colorado River water between the lower basin states

(California, Arizona, and Nevada), which sets the stage for much larger reallocations of water in the desert Southwest. Among other things, the new federal rules will allow Arizona to store its unused Colorado River water in underground aquifers and then sell that water to other states. Since California has been getting Arizona's surplus for free, the state will have to search hard to make up for these lost supplies. IID's agricultural water will look more and more attractive.[22]

As with pricing options, a range of water marketing possibilities exist. Farmers within irrigation districts can trade water on a short-term basis in response to conditions that year, as they are doing in the Northern Colorado and Westlands districts. They can also enter into long-term contracts—5, 10, or even 30 years—to supply a certain amount of water to a willing buyer while retaining the rights to that water. IID's deals with southern California municipal water agencies are of this type. In addition, farmers can sell their water rights to another user, in which case the legal water entitlement permanently changes hands.

A fourth option, and one that deserves more attention, is trading through a water bank. A farmer (or other water user) "deposits" water in the bank in hopes that another user will rent that water at a price that yields a profit for the depositer. The "bank" is run by a regional water authority, a state agency, or some other entity that can effectively manage the process, including setting the rental terms. Water districts that allow their irrigators to buy and sell water among each other, such as the Northern Colorado Water Conservancy District, are in effect running a water bank.

Banks could play a bigger role in reallocating water than they have to date, however, because they provide an efficient way to transfer water among users in different systems as well as between farms and cities. The beauty of a water bank is that it gives flexibility to farmers who often face great uncertainty about their water needs in a particular year. They can choose to deposit or rent water one year, but then do something different the next.

For example, if crop prices drop and it becomes uneconomical to irrigate, they can earn some cash by banking their water temporarily rather then being pressured into a longer-term sale. Likewise, if a drought strikes, they can choose to rent water if they have perennial crops that need an additional irrigation or two, or they might leave some land fallow and increase their income by selling water to others willing to pay more for it. In this way, a water bank can help farmers weather rough years and stay in business. Water banks can also help whole regions cope with drought.

In the western United States, two large operations give a hint of water banking's potential. Idaho established a water bank in 1979. Administered by the state's Water Resource Board, the bank accepts either permanent or temporary entitlements for sale or lease to willing buyers. Water transactions typically total several hundred thousand acre-feet a year. California instituted a water bank in 1991, during the fourth consecutive year of a severe drought. Operated by the state's Department of Water Resources, the bank purchased 820,000 acre-feet (about 1 billion cubic meters) of water that year, half of it coming from irrigators who had agreed to stop irrigating crops for part of the year. The buyers paid $175 per acre-foot, and included cities, farmers, and the state-run water supply project. State officials put the drought bank into effect again in 1992 and in 1994, but much less water changed hands then. California plans to operate the bank whenever drought or scarcity conditions create critical shortages.[23]

Chile and Mexico are among the few developing countries that are actively promoting water marketing. Both countries have put in place a system of legally defined water rights. In most developing countries, however, farmers do not have secure ownership of water, which prevents them from entering into agreements to sell it. Many of these farmers do, however, trade water informally through spontaneous spot markets. A 1990 survey of surface canal systems in Pakistan found active water trading in 70 percent of them. In *warabandi* systems, trades might simply consist of one farmer relinquishing a scheduled turn to a neighboring farmer.[24]

Groundwater is also actively traded, especially in South Asia. Ruth Meinzen-Dick, a research fellow with the Washington-based International Food Policy Research Institute, finds that one fifth of all well owners in Pakistan sell some of their water, and that informal groundwater markets are operating in every province. Typically, smaller landowners and tenant farmers who cannot afford their own tubewells purchase water from a larger or wealthier landowner. They may pay an hourly fee that covers the cost of the fuel plus wear and tear on the pump, or return to the well owner a share of the crops they produce. Although farmers purchasing water have less control over their irrigation supplies than the well owners have, they are still able to profit from buying the water, since it raises their yields and increases their incomes.[25]

Indian economist Tushaar Shah has a similarly positive view of the informal groundwater markets that have sprung up in parts of India. He estimates that as much as half of the land currently irrigated by private wells and pumps in India belongs to buyers of water. Most of the markets are limited to a village or part of a village, and, as in Pakistan, they arise spontaneously in response to the mutual gain potential buyers and sellers can realize. When Shah and colleague Vishwa Ballabh took a close look at water markets in six villages in north Bihar, they found, surprisingly, that buyers of water had cropping intensities and crop yields equal to or greater than those of the well owners. The markets effectively provided small farmers with water supplies that were timely, reliable, and sufficient to achieve productivity levels that rivaled farmers who owned their own wells.[26]

On the downside, Shah and Ballabh found considerable

evidence of monopolistic pricing. Where competition among sellers was lacking, well owners would charge two to three times as much as their incremental pumping costs, pocketing substantial extra profits. While acknowledging that this gouging of poorer farmers needs to be addressed, Shah and Ballabh are nevertheless enthusiastic about water marketing's role. Water markets have taken shape, they conclude, "as a robust and dominant irrigation institution serving as virtually the sole powerhouse energising north Bihar's new-found agrarian dynamism."[27]

Informal water markets are good examples of bottom-up institution building. They have emerged to fill an institutional vacuum—the absence of practical rules and operations for allocating water among farmers. While they are functioning remarkably well in many places, some intervention is needed to ensure that monopolistic water lords do not charge excessively for their water, that groundwater pumping does not exceed natural recharge, and that large well owners do not draw water tables below the level that small farmers with only manual pumps can reach. Adam Smith's invisible hand may get the water markets going, but some rules of the game are needed to ensure fair play.

Giving Farmers More Control—and Responsibility

As in electrification, transportation, and other capital-intensive sectors of the economy, building and maintaining large irrigation networks has greatly stressed the budgets of governments large and small. Bloated bureaucracies, lavish subsidies, blatant rent-seeking, and all manner of perverse incentives have led to not only inefficiency and lack of accountability but also financial unsustainability. Many governments simply cannot afford to foot so large a bill for irrigation much longer.

Whether out of desperation or opportunity, many governments are putting irrigation on the bandwagon of privatiza-

tion—transferring responsibility for and sometimes ownership of public assets to the private sector. More than 25 countries are now in the process of turning irrigation systems over
to farmers' groups or other private organizations. This institutional shift goes by different names in different places—self-
management in Niger, participatory management in India,
privatization in Bangladesh, turnover in Indonesia and the
Philippines, and management transfer in Mexico and
Turkey—but in all cases it represents a devolution of authority and responsibility from governments to irrigators.[28]

The Philippines was one of the first nations to go down this
path. For centuries, community farmers' groups there built,
managed, and maintained their own irrigation systems. After
1950, however, when irrigation worldwide began to spread
rapidly, the national government assumed responsibility for
building large irrigation projects and took control of many of
the older community systems as well. During the early 1960s,
Philippines officials created the National Irrigation Administration (NIA) as a semiautonomous corporation with authority and funding to build irrigation systems. Like many
irrigation bureaucracies, the NIA did not get to keep the fees
collected from farmers; rather, the money went into the
national treasury. Over time, as the share of collectible fees
actually paid decreased and as NIA's budget deficits rose, officials realized that something had to be done.[29]

In 1974, the government revised NIA's operating rules,
allowing the agency to keep the irrigation fees it collected. The
government planned to phase out, over five years, its subsidy
for operation and maintenance of irrigation systems. With the
subsidies removed, NIA would be wholly reliant on the fees
paid by farmers.[30]

Although hardly revolutionary, these simple rule changes
dramatically shifted the incentives driving all the players' decisions. NIA had obvious reasons to collect fees, because these
funds became the sole source of its income. Farmers had rea-

son to pay the fees, because their money went toward maintaining and upgrading their irrigation networks. They became customers paying for a service, rather than beneficiaries of government handouts. For its part, the government had reason to see that the reforms worked in order to get some relief from mounting budgetary pressures.

In addition to the financing changes, NIA gradually turned responsibility for irrigation systems over to the farmers. This has required the strengthening of numerous farmers' associations, a vital aspect of reform that was not easy to carry out. As Elinor Ostrom tells it in her book, *Crafting Institutions for Self-Governing Irrigation Systems*, "It took the creative energies of many inspired public officials, newly hired irrigation organizers, and devoted academics and solid support from the Ford Foundation to organize strong user associations that could relate effectively with an all-powerful supplier like the NIA." Over the longer term, the government hopes to fully privatize small and medium irrigation schemes, and to limit NIA's role to large central features of the water delivery system.[31]

The hard work of irrigation reform in the Philippines appears to be paying off. Studies of an irrigation system in southern Luzon, for example, found that after responsibilities were turned over to farmers, fee collection rose from 20 percent to more than 80 percent. Farmers managed to cut the cost of running the system by a fourth. The system's annual budget deficit nearly disappeared. In addition, the number of NIA staff assigned to this project area dropped by 75 percent, trimming the national irrigation bureaucracy.[32]

More important, studies by the Institute of Philippine Culture show that when irrigators participate in the design, construction, and subsequent management of irrigation systems, agricultural productivity increases. The institute examined 46 systems that NIA was upgrading during the early 1980s, 24 of which included participation by irrigators. The researchers found that in systems with farmer participation, dry-season irrigation rose by 35 percent, compared with 18 percent in those without farmer participation. Even more compelling, rice yields of the participatory systems rose an average of 21 percent during the dry season, while those of the nonpartici-patory systems fell by 1 percent. Systems were better built when the local knowledge and concerns of farmers were included from the outset. Farmer participation also helped ensure that scarce water was allocated fairly and effectively.[33]

Mexico is currently engaged in the largest and fastest turnover of irrigation systems anywhere in the world. Follow-ing the Mexican revolution, the new government nationalized water resources and took over responsibility for irrigation sys-tems. It invested heavily in expanding irrigation to foster grain self-sufficiency and promote agricultural exports. By 1982, irrigated land nationwide stood at 5.3 million hectares.[34]

That year, however, the Mexican peso collapsed, and the resulting financial crisis stopped irrigation expansion in its tracks. Government spending on the program plummeted from $3.6 billion in 1981 to $230 million in 1990. Water fees had not kept pace with rising costs, so funds for operation and maintenance dwindled. Irrigation fees covered more than 85 percent of operation and maintenance costs in the early 1950s, but less than 20 percent by the early 1980s. Not surprisingly, Mexico's irrigation infrastructure began to deteriorate badly. By the end of the 1980s, 800,000 hectares of irrigated land were yielding only a small fraction of their potential or were out of production altogether. An additional 1.5 million hectares needed rehabilitation work to prevent deterioration. This required money that the government did not have and that international lenders, grappling with the Latin American debt crisis, were reluctant or unable to provide.[35]

As in the Philippines, the Mexican government came up with a plan for making publicly provided irrigation financially more self-sufficient, and it involved decentralizing and devol-ving responsibility to the irrigators themselves. Begun in the early 1990s, the transfer program gives water user associations responsibility for operating and maintaining irrigation systems within their service area, as well as for collecting fees to cover costs. A key goal of the program is to eliminate government subsidies to these districts altogether. The water user associations are legal entities under Mexican law, and they enter into fixed-period contracts, typically 5 to 50 years, with the government. Instead of rights to a specific amount of water, each irrigation association has rights to a fixed proportion of the supply available that season, with respective shares based on the irrigated area under the association's domain.[36]

By June 1996 the government had turned management over to irrigation associations on 2.8 million hectares of irrigated land—more than 85 percent of the 3.3 million hectares of publicly irrigated land. As water fees rose to cover costs, the irrigation districts went from 37 percent financial self-sufficiency to 80 percent. The cost of water to the farmers, although higher, generally remained within 3–8 percent of their total production costs, a typical range for irrigated agriculture. The number of employees engaged in irrigation operations in the national water agency dropped by 42 percent, further cutting government expenditures.[37]

Maintenance improved too, at least in some areas. A study of one irrigation district showed that the upkeep of gates, canal roads, and other structures improved, and the volume of sediment removed from canals and drains increased fivefold. Nearly two thirds of this district's farmers considered maintenance to be poor before the government turned the system over but good afterward.[38]

Mexico's newly strengthened irrigation associations have yet to build up reserve funds to deal with emergencies. But the whole irrigation base is now much more financially stable. During the fiscal crisis of 1995, when the government provided almost no operating funds to its line agencies, the irrigation districts had ample money from fees to keep operations running smoothly. It is too early to judge other important criteria for the success of Mexico's program—in particular, whether it promotes greater equity and fosters more-efficient use of scarce water. If imitation is the best form of flattery, however, Mexico has reason to boast: water officials from Turkey were sufficiently impressed with what they saw during visits to Mexico that they returned home to initiate a similar program.[39]

As the pendulum swings from state control of irrigation to private control, it is easy to overlook a promising set of options in between. Many well-functioning irrigation systems are run communally, with rights and responsibilities determined by the farmers themselves. These autonomous, self-governing, communal arrangements present a third way of allocating and managing irrigation water—one that offers the cost efficiency and customized design of privately run systems, but at the same time can be grounded in principles of equity and responsibility.

Communal systems treat irrigation water as the common property of a group of farmers—those tied to a village, town, or some other defined entity. Western market economies generally look askance at common property management because of perceived conflicts between individual and community interests. Ecologist Garrett Hardin, in his influential 1968 essay, "The Tragedy of the Commons," argues that each person acting out of self-interest will lead to collective action that destroys the common resource. "Freedom in a commons," he wrote, "brings ruin to all."[40]

Although it is easy to find examples to support Hardin's view, his argument blurs an important distinction. Just because a resource is the property of a community rather than an individual does not mean that people necessarily have free and unregulated access to that resource. In the management of water, as with forests, fisheries, and grazing lands, there are many examples of successful and long-lasting community-based arrangements in which a set of rules and traditions established by the users themselves regulates access to a common resource.

In the northern Philippines, for example, landless laborers organized themselves into systems called *zanjeras* with the aim of acquiring rights to use unirrigated land owned by wealthier farmers. The group members work together to build a canal that supplies water to a portion of the landowner's fields in exchange for rights to farm that land. Each person who chooses to participate is given some land to cultivate, but in return the group expects that person to help construct and maintain the irrigation system. Rights and responsibilities are bundled together so that benefits are equated with costs. Because each participant's land is divided into three parcels—one each in the head, middle, and tail portions of the canal system—water is allocated equitably. The tail-end inequities so typical in government-run schemes are virtually nonexistent.[41]

The Subaks of Bali, Indonesia, offer another good example. Like the *zanjeras*, these farmers' associations constructed their own irrigation systems with very little outside assistance. Their rules specify duties and obligations, which include contributions of cash and labor to operate and maintain the system. Water is allocated according to a principle called *tektek*, which gives each member rights to a proportion of the available supply. That proportion is determined not just by how much land the farmer has, but also by soil conditions, the farmer's role in the association, the farmer's initial investment in the system, and other factors. Because the communities are relatively stable and migration is low, participating farmers take a long-term view, which enhances the benefits of cooperating. Long-standing village and religious traditions support the activities of the irrigation associations. Priests, for example, play an active role in allocating the water supply, lending legitimacy to the process. As in the *zanjeras*, there is a strong emphasis on fairness.[42]

The beauty of self-governance is that it allows farmer groups to customize irrigation operations to suit their specific physical and cultural conditions. In contrast, government bureaucracies often impose uniform solutions on very different circumstances. Just like private irrigators, however, those in self-governing communal arrangements need long-term security in their water rights if they are to invest in their systems. Unless the prevailing rules of water allocation in any society acknowledge the validity of informal customary water rights, these communal arrangements are at risk of losing their water to those with formal legal entitlements—even if the customary rights predate the legal rights.[43]

A Missing Piece—Groundwater Rules

If the trends described in the last few sections accurately reflect the way irrigation will evolve in the years ahead, a number of changes seem clear. First, governments will end up with a smaller management role and farmers themselves with a larger one. Second, more irrigators will play a proactive part in reallocating water—both among themselves and to cities—as water markets and water banks open up and spread. And third, many if not most farmers will pay more for water as they begin to bear a greater share of irrigation's costs.

If irrigation's software designers do their jobs well, these new rules of the game will make irrigated agriculture more efficient, equitable, and financially sound. But one critical piece of institutional design has so far received strikingly little attention: curbing the overuse of groundwater.

Depletion of groundwater looms as one of the biggest threats to irrigated agriculture. Numerous countries with large populations and vast irrigated areas are drawing down aquifers to meet today's needs, leaving less to meet tomorrow's. With groundwater, the tragedy of the commons is playing out full tilt. In most places, any farmer who can afford to sink a well and pump water to the surface can do so unrestrained. Ownership of land typically implies the right to water pumped from beneath that land. As the number of private wells exploded during recent decades, irrigators collectively began to extract more water than nature was able to replenish. If unchecked, the degree of overpumping documented in Chapter 4 leaves the world vulnerable to sudden and dramatic cutbacks in food production.

No government has made a concerted effort to solve the problem of groundwater overpumping. Officials in India circulated a "model groundwater bill" in 1992, but none of the states have passed legislation along those lines. Some have made efforts to regulate groundwater use through licensing, credit, or electricity restrictions, or by setting minimum well-spacing requirements. But no serious efforts have been made to control the volume of water extracted. V. Narain of the Tata Energy Research Institute points out that "groundwater is viewed essentially as a chattel attached to land," and there is "no limit on how much water a land-owner may draw."[44]

Indian researchers and policymakers broadly agree that rights to land and water need to be separated. Some have argued for turning de facto private groundwater rights into legal common property rights conferred upon communities in a watershed. But instituting such a reform is extremely difficult. Wealthy farmers, who have the ear of politicians on this issue, do not want to lose their ability to tap groundwater on their property in any quantity they desire.[45]

The United States has no official national groundwater policy, and, as in India, it is up to the states to manage their own groundwater resources. So far, only Arizona has passed a comprehensive groundwater law that explicitly calls for balancing groundwater use with recharge. Arizona's strategy for meeting this goal by 2025 involves substituting Colorado River water imported through an expensive federally subsidized project for some of its overpumped groundwater. But few regions can rely on such an option, which merely replaces one type of excessive water use with another.

Virtually everywhere, governments and farmers have their heads in the sand on the groundwater problem—but it is not going away. Irrigation cutbacks will occur. The only question is whether they occur in a planned and orderly fashion, or abruptly and chaotically.

A first step toward taking charge of the problem is for governments to commission credible and unbiased assessments of the long-term rate of recharge for every groundwater basin or aquifer. This would establish the limit of sustainable groundwater use. The second step is for all those involved—including scientists, farmers' groups, and government agencies—to devise a plan for balancing pumping with recharge levels. If current pumping levels exceed sustainable limits, achieving this goal will typically involve some mix of pumping reductions and artificial recharge—the process of channeling rainfall or surplus river water into the underground aquifer.

Arriving at an equitable way of allocating groundwater rights such that total pumping remains within sustainable levels will not be easy. Legislatures or the courts might need to invoke a legal principle that elevates the public interest over private rights. One possibility, for example, is the public trust doctrine, which asserts that governments hold certain rights in trust for the public and can take action to protect those rights from private interests. Some scholars have recommended its use to deal with India's groundwater problem. Recent rulings in the United States show that this legal instrument is potentially very powerful. The California Supreme Court ordered Los Angeles to cut back its rightful diversions of water from tributaries that feed Mono Lake, declaring that the state holds the lake in trust for the people and is obligated to protect it. A broad interpretation of the public trust doctrine could have sweeping effects, since even existing rights can be revoked in order to prevent violation of the public trust.[46]

Establishing a legal basis for limiting groundwater use is only the beginning. Many additional hurdles block the path of devising a practical plan for making groundwater use sustainable. Mexico is one of the few countries that seems to be tackling the task head on. Following the enactment of a revised water law in 1992, Mexico created broadly participatory River Basin Councils. The council for the Lerma–Chapala River basin, for example, an area that contains 12 percent of Mexico's irrigated land, is in the process of setting specific regulations for each groundwater aquifer in the region. Technical committees charged with devising plans to reduce groundwater over-pumping include a broad mix of players, including the groundwater users themselves, lending legitimacy to both the process and the outcome.[47]

Governments and other authorities can also ease the pain of pumping cutbacks and minimize crop production losses by offering carrots along with the sticks. In the United States, the High Plains Underground Water Conservation District in Lubbock, Texas has worked closely with farmers to design and spread efficient irrigation technologies and management practices in order to slow the depletion of the Ogallala aquifer. Despite lax regulations on groundwater use in Texas, as in most parts of the United States, this partnership between the groundwater district and the farmers has produced very positive results. (See Chapter 8.)

=========

Most of the hard work of crafting new irrigation software has yet to be done. The design of workable public, private, and community-based arrangements must happen locally and regionally, and the outcome will vary from place to place. Whether these new arrangements improve irrigated agriculture's performance and sustainability—the bottom line—remains to be seen. The various experiments under way deserve careful scrutiny as we search for rules of the game that can make irrigation more efficient, equitable, and ecologically sound in the twenty-first century.

11

LISTENING TO
OZYMANDIAS

And on the pedestal these words appear:
"My name is Ozymandias, King of Kings:
Look upon my works, ye Mighty, and despair!"
Nothing besides remains. Round the decay
Of that colossal wreck, boundless and bare
The lone and level sands stretch far away.

Percy Bysshe Shelley
Ozymandias

History tells us that most irrigation-based societies fail. The quest for greater fertility and productivity can lead to infertility and decline if pursued too vigorously, in the wrong places, or in ways that defy basic ecological principles. This lesson creates for us an uncomfortable conundrum, because irrigation is more essential than ever. Some 40 percent of our food comes from irrigated lands, and we are betting on that share to increase in the decades ahead. Coaxing more food from a portion of our cropland allows a larger area of forest and grassland to remain wild—free to perform the work of nature that we and all life depend on. We return to the questions posed at the outset of this book: Can our modern irrigation society succeed? What will success require?

The parallels between the state of irrigated agriculture

today and the problems that plagued failed irrigation societies of the past are both striking and unsettling. From the buildup of salts in fertile soils to the flaring of tensions over disputed waters, many of today's threats echo those of the past. But whole new elements in the irrigation drama are unfolding today that present a more difficult set of challenges than ever.

First, signs of trouble are surfacing with unprecedented speed. Sixty percent of our irrigation base is less than 50 years old, yet threats to the continued productivity of much of this land are already apparent. A great deal of this newer irrigated land depends on groundwater, which farmers began tapping on a large scale only after 1950, as powerful diesel and electric pumps became accessible. Today the problem of overpumping—extracting more water from underground reservoirs than nature puts in—is pervasive in most major areas of irrigated agriculture. Collectively, farmers are racking up an annual water deficit of at least 160 billion cubic meters—enough to produce half the U.S. grain harvest. This deficit suggests that a tenth of the world's current grain supply is propped up by unsustainable water use, a situation that cannot persist indefinitely.

Second, irrigation's land and water base are rapidly being eroded by salinization of soils, siltation of reservoirs, and the siphoning off of irrigation water to supply burgeoning cities. Official statistics still suggest that the world's irrigated area is expanding, but net growth actually might be close to zero if the negative side of the ledger were counted properly.

Grave as these developments are, political leaders have scarcely taken notice of them—a situation ominously reminiscent of that in ancient Mesopotamia. Accounts of the final stages of Mesopotamia's decline point to serious problems of waterlogging and salinization from the expansion and intensification of crop production. Scholar Peter Christensen notes, however, that Mesopotamian rulers rarely mentioned these problems in their records. "Perhaps," he muses, "the processes were too gradual to attract attention; more likely, the chroniclers were simply not aware of the importance of these things."[1]

Third, the phenomenal successes of the Green Revolution have left a legacy of complacency that belies the scale of the challenge ahead. Merely holding on to the agricultural gains of the last 50 years will be no small feat, because a good share of them were won through ecologically unsound uses of land, water, and chemicals. Short-run but ultimately unsustainable production gains are particularly vexing, because they invite a degree of comfort that is wholly unwarranted.

Again, the story of Mesopotamia is instructive. Large-scale irrigation in the plains between the Tigris and Euphrates Rivers proved productive enough in the short term to support flourishing, vibrant, and amazingly inventive societies. But as Jared Diamond puts it bluntly in his Pulitzer Prize–winning book, *Guns, Germs, and Steel*, the societies of the Fertile Crescent "committed ecological suicide by destroying their own resource base." Diamond refers not only to salinization of irrigated lands but also to deforestation of the Mesopotamian highlands and the consequent erosion of hillsides and siltation of valley lands. Short-term success does not secure long-term prosperity.[2]

As if maintaining the hard-won advances of the last few decades were not difficult enough, we now face the need for a repeat performance. The most likely scenario in recent U.N. population projections suggests that an additional 3 billion people will join humanity's ranks during the first half of the twenty-first century. Some 3.5 billion people were added during the last half of the twentieth century. So farmers, scientists, and engineers are called upon to do again what they did during the last 50 years—but this time with a greatly depleted larder of natural resources.[3]

Water scarcity, in particular, will make a second round of success extremely difficult, as earlier chapters have demonstrated. In addition to depleting groundwater, the scale of water extractions today is sapping the health of the aquatic environment. Additional pressures on rivers, lakes, and wetlands will destroy more fisheries, consign more species to extinction, and unravel more of the ecological services that support the human economy. One research team has placed the value of wetlands alone at $4.9 trillion per year—equivalent to one eighth of the current gross world product. Removing more nuts and bolts from such valuable Earth machinery is risky, to say the least. But this is precisely what happens each time we build an additional dam, river diversion, or other piece of engineering that dismantles ecosystems or disrupts their functions.[4]

Fourth, the scale of the challenge ahead is also unprecedented because of the rapidly growing imbalance between human numbers and Earth's distribution of fresh water. Over the next quarter-century, as noted in Chapter 6, we will see more than a sixfold increase in the number of people living in water-stressed countries—from 470 million today to 3 billion in 2025. Water-stressed countries have difficulty mobilizing enough water to satisfy all the food, household, and material needs of their populations. As a result, most will choose to shift water out of agriculture to other uses that create more economic value. For example, 1,000 tons of water can be used to produce one ton of grain, which has a market value of roughly $200, or industrial products worth $10,000–20,000. Water used in industry also creates more jobs than the same volume used in farming, increasing pressures to shift scarce supplies away from agriculture.

As the number of people in water-stressed countries climbs toward 3 billion, competition for water will spread across borders through the global grain trade as more countries attempt to import enough grain to fill their food gaps. Whether the United States, Western Europe, and other food exporters will be able to satisfy these demands is only half the issue. Equally important is whether the importers—primarily poor nations of South Asia and sub-Saharan Africa—can afford to buy the grain they need. No global food models adequately incorporate water scarcity into projections of exportable food surpluses, future food prices, or access to food by the poor. As a result, they present too sanguine a picture of future food security and perpetuate unwarranted complacency.

Last, climate change on the scale that scientists are projecting for the next century adds a whole new dimension to the food and water challenge. A global temperature shift of the magnitude predicted has not occurred since the dawn of settled agriculture. The relative stability of Earth's climate over the last 10,000 years is an anomaly in recent geologic history, albeit a fortuitous one for us. As the most weather-dependent of activities, agriculture almost certainly advanced more rapidly than would have been the case had large and unexpected temperature and rainfall shifts occurred.

History shows that climate wild cards—especially droughts, floods, and other events that alter rainfall and river flows—can overwhelm a seemingly advanced society's ability to cope. As described in Chapter 2, archeological evidence suggests that the abandonment of the Akkadian city of Tell Leilan in Mesopotamia and the sudden disappearance of the Hohokam from the American Southwest may both be linked to unexpected changes in water availability. In the first case, drought caused a mass migration southward of people from the rain-fed plains to the irrigated valley lands of the Euphrates River. The valley's food-production system could not support such an influx, especially with the flow of the Euphrates greatly diminished by the drought. The cause of the Hohokam's vanishing is more speculative, but researchers find evidence that episodes of both drought and flooding may have played a part.

Any one of the stresses evident today—salinization, siltation, faltering aquatic ecosystems, mounting competition for water, the growing imbalance between populations and available water supplies, and global-scale climate change—would seriously challenge irrigated agriculture's future productivity. But these stresses are evolving simultaneously, which magnifies the constraints on future food production and heightens the risk that rising food prices or larger pockets of hunger will destabilize civil societies.

The best antidotes to stress—both for an individual and for society at large—are preparedness and flexibility. At the moment, we do not have enough of either trait to see us safely through the complex of stresses on the horizon. As described in earlier chapters, irrigated agriculture is unprepared for the era of water scarcity and competition that is rapidly emerging. Virtually all of the most promising techniques for improving water productivity—from drip irrigation to the use of computerized weather and soil monitoring—remain vastly underused because water pricing and other "rules of the game" do not encourage their adoption. Rigid water institutions—including policies that preclude the trading of water—hinder the flexibility so essential to responding to times of stress. And the lack of laws and policies to confront groundwater overpumping, depleted river flows, and other environmental damage leaves irrigated agriculture vulnerable to sudden cutbacks and robs the next generation of its productive potential.

Smarter policies and better technologies thus have an important role to play in working toward the goal of enough food and water for all in the next century. As Chapter 9 points out, part of the needed "technical fix" is providing poor farmers with access to affordable small-scale irrigation technologies that enable them to lift their food production and incomes directly—thereby eliminating large pockets of chronic hunger. Untapped potential also lies with new crop varieties that can better withstand salt and water stress, the production of salt-loving plants with seawater irrigation in coastal deserts, and strategic ways of better matching water of varying qualities with different uses.

Taken together, these kinds of efficient technologies and new agronomic measures can move us a long way toward the goal of doubling water productivity. But successfully meeting the challenge ahead will also require a scaling back of consumption so as to reduce our individual and collective pressures on the planet—in effect, cutting some slack in a system that will otherwise be on the verge of snapping.

In his thoughtful book *The Ecology of Eden*, Evan Eisenberg describes the notion of *tsimtsum*—a voluntary retraction to create space for other living things. As Eisenberg tells it, the term comes from a sixteenth-century kabbalist teaching that in order to create the world, God—who was everywhere—had to "draw himself inward—to take a step back, as it were... in order to leave a space where other things could exist." This self-retraction was called *tsimtsum*.[5]

We may need just such a collective pulling back of our human presence in order to create room for additional people, for other living things, and for the work of nature to continue. Because of the size of the human population, even small changes in individual consumption can have a large collective impact on the environment. By shifting dietary choices, for example, individual consumers can practice *tsimtsum*. The typical American diet, with its large share of animal products, requires twice as much water to produce as nutritious but less meat-intensive diets that are common in developing countries and some Asian and European nations. By moving down the food chain, Americans could get twice as much nutritional benefit out of each liter of water consumed in food production. Stated otherwise, the same volume of water could feed two people instead of one, leaving additional water in rivers and streams to help restore fisheries, wetlands, recreational opportunities, and ecological functions overall.[6]

It may be only a matter of time before China restrains the consumption of pork, which on a per capita basis now matches that in the United States. Producing pork takes roughly twice as much grain per kilogram as producing chicken or farm-raised fish does. In a country with 21 percent of the world's population but only 7 percent of the renewable fresh water, China may have little choice but to encourage less grain-intensive and more water-efficient diets as portions of its population become more affluent.[7]

Lessening the water demands of cities and industries is also essential. Besides overtaxing and polluting rivers, lakes, and aquifers, these demands will partially be met by siphoning water away from agriculture. Numerous opportunities exist to improve water productivity in homes, offices, commercial enterprises, and manufacturing facilities—and this potential has barely been tapped. We have grown accustomed to measuring the energy intensity of our economies during the last quarter-century, but few countries have even begun to track their water intensity.

It is difficult to imagine how we can succeed in meeting human needs sustainably without the adoption of a guiding water ethic grounded in the principles of sufficiency and sharing. Water has two fundamental traits that distinguish it from any other resource. First, it is a prerequisite for life—no plant or animal, humans included, can survive without it. Second, there are no substitutes for water in most of its uses. Together, these attributes lend an overarching ethical dimension to our individual and collective decisions about water—one that says enough water should be provided to sustain all people and living systems before some get more than enough. This is not a pie-in-the-sky prescription, but increasingly the only realistic remedy to meeting the challenges at hand.

I could not help but realize, in a visceral way, the need for such an ethic as I stood amid the dessicated landscape of the Colorado River delta in northern Mexico in 1996. Around me was a small community of Cocopa Indians whose ancestors had fished and farmed in the delta for more than a thousand years. Their culture is on the brink of extinction because too little river water now makes it to the delta to sustain them. Along with the endangered Cocopa, two endangered species— the Yuma clapper rail and the desert pupfish—hang on in the delta by way of small pockets of wetlands that are anything but secure homes. Gone are the vast flocks of waterfowl and abundant wildlife that American naturalist Aldo Leopold recorded seeing in this "milk-and-honey wilderness" in 1922.[8]

What did we gain when we traded away these cultural and biological values, driving long-settled people from their homes and wildlife from their habitats? The answer seems to be more swimming pools in Los Angeles, more golf courses in Arizona, bigger oasis cities in the American Southwest, and more desert agriculture. This tradeoff increased water's contribution to the U.S. gross national product, but at the uncounted cost of lost biological and cultural diversity and of irreplaceable natural assets. Had a guiding ethic of sufficiency and sharing been adhered to when Colorado River water was allocated, enough would have been provided to sustain the Cocopa people and the ecosystems of the Colorado delta before water went to luxury activities upstream.

In short, making irrigated agriculture both productive and sustainable is part of the larger challenge of restoring balance between the demands of the human population and the biological requirements of Earth's living systems. Humanity now appropriates for its own use more than half of Earth's accessible renewable fresh water and some 40 percent of its net photosynthetic product. This degree of human dominance leaves a dangerously thin margin of support for the millions of other species with which we share the planet—species that perform the vital work of nature on which our societies rest. It is not enough to meet a short-term goal of feeding the global population. If we do so by consuming so much land and water that ecosystems cease to function, we will have not a claim to victory but a recipe for economic and social decline.[9]

Population and consumption growth each have the ability to overwhelm any gains we might make from technological advances. Reducing both sources of rising human pressures forms the core of the collective pulling back, or *tsimtsum*, that is needed. Although the latest U.N. projections show a middle-level trajectory that brings global population to 8.9 billion in 2050, the low-growth projection has population peaking at 7.5 billion in 2040 and declining slowly thereafter. A move from the medium to the low-growth path would provide a great deal of additional breathing room and flexibility with which to achieve the needed balance between humans and nature.[10]

Like a weary marathon runner who gets a welcome surge of energy toward the end of the race, perhaps we can muster more vigorous work and investment to slow population growth further with the realization that the goal of stabilizing human numbers is within sight. Stepped-up support for groups and agencies dedicated to improving family planning services, women's reproductive health, and educational and economic opportunities for women will better women's lives as well as lower birth rates. Coupled with reductions in high-end consumption, slower population growth would make the goal of adequately feeding all people while keeping ecosystems intact—as well as the creation of a sustainable society overall—far more achievable.

It is possible to envision a modern form of irrigated agriculture that is resource-renewing rather than resource-depleting and that preserves Earth's natural capital rather than liquidates it. Just as the application of hydraulic principles coupled with new engineering capabilities reshaped the face of irrigated agriculture during the late nineteenth century, so the application of ecological principles combined with new information and irrigation technologies can guide its redesign in the twenty-first century. The less pressure we place on our agricultural base, the more likely we are to succeed.

Time, however, is of the essence. These transformations cannot happen overnight, and so it is important that we begin the process now. If we fail to make a timely transition, we will see—just as our predecessors in the Fertile Crescent did—that we do indeed rest on a pillar of sand.

NOTES

Chapter I. New Light on an Old Debate

1. Irrigated area in 1800 from K.K. Framji and I.K. Mahajan, *Irrigation and Drainage in the World* (New Delhi, India: Caxton Press Private Ltd., 1969); U.N. Food and Agriculture Organization (FAO), *1996 Production Yearbook* (Rome: 1997); for views of agricultural experts, see David Seckler, David Molden, and Randolph Barker, *Water Scarcity in the Twenty-First Century*, Water Brief No. 1 (Colombo, Sri Lanka: International Water Management Institute, 1998); Daniel Hillel, presentation at The Keystone Center Workshop on Critical Variables and Long-Term Projections for Sustainable Global Food Security, Airlie House, Warrenton, VA, 10–13 March 1997.
2. Sandra Postel, *Dividing the Waters: Food Security, Ecosystem Health, and the New Politics of Scarcity*, Worldwatch Paper 132 (Washington, DC: Worldwatch Institute, September 1996).
3. Figure of 1 billion from Seckler, Molden, and Barker, op. cit. note 1; Sandra Postel, "Water for Food Production: Will There be Enough in 2025?" *BioScience*, August 1998.
4. The Keystone Center Workshop on Critical Variables and Long-Term Projections for Sustainable Global Food Security, Airlie House, Warrenton, VA, 10–13 March 1997.

5. U. S. Department of Agriculture (USDA), *Production, Supply, and Distribution*, electronic database, Washington, DC, updated February 1999.

6. Known as Malthus's first essay, since it was later revised, the full title was "An Essay on the Principle of Population as It Affects the Future Improvement of Society"; see Thomas Robert Malthus, *An Essay on the Principle of Population* (1798), in Philip Appleman, ed., *The Norton Critical Edition* (New York: W.W. Norton & Company, 1976).

7. Crookes cited in Tim Dyson, *Population and Food: Global Trends and Future Prospects* (London: Routledge, 1996).

8. Joel E. Cohen, *How Many People Can the Earth Support?* (New York: W.W. Norton & Company, 1995); post-1950 figures from U.S. Bureau of the Census, *International Data Base*, electronic database, Suitland, MD, updated 30 November 1998.

9. L.R. Oldeman, V.W.P. van Engelen, and J.H.M. Pulles, "The Extent of Human-Induced Soil Degradation," Annex 5 of L.R. Oldeman, R.T.A. Hakkeling, and W.G. Sombroek, *World Map of the Status of Human-Induced Soil Degradation: An Exploratory Note* (Wageningen, Netherlands: International Soil Reference and Information Centre, 1991).

10. Status of major fishing areas from Maurizio Perotti, fishery statistician, Fishery Information, Data and Statistics Unit, Fisheries Department, FAO, Rome, e-mail to Worldwatch, 14 October 1997; per capita from Anne Platt McGinn, "Fisheries Falter," in Lester R. Brown, Michael Renner, and Brian Halweil, *Vital Signs 1999* (New York: W.W. Norton & Company, 1999).

11. Cropland loss to erosion from G. Leach, *Global Land and Food in the 21st Century: Trends & Issues for Sustainability* (Stockholm, Sweden: Stockholm Environment Institute, 1995), and from David Pimentel et al., "Environmental and Economic Costs of Soil Erosion and Conservation Benefits," *Science*, 24 February 1995; FAO, *1995 Production Yearbook* (Rome: 1996).

12. USDA, op. cit. note 5; USDA, *Grain: World Markets and Trade*, December 1996 and January 1997.

13. Source of water for crops from Postel, op. cit. note 3; the IMPACT model of the International Food Policy Research Institute (IFPRI) projects annual growth in cereal yields in industrial countries of less than 1 percent per year from 1993–2020, see Per Pinstrup-Andersen, Rajul Pandya-Lorch, and Mark W. Rosegrant, *The World Food Situation: Recent Developments, Emerging Issues, and Long-Term Prospects* (Washington, DC: IFPRI, 1997); for a discussion of yield trends and potentials, see also Lester R. Brown, *The Agricultural Link: How Environmental Deterioration Could Disrupt Economic Progress*, Worldwatch Paper 136

(Washington, DC: Worldwatch Institute, August 1997).

14. Number malnourished from FAO, *The Sixth World Food Survey* (Rome: 1996).

15. Capital value calculated by multiplying the 1995 global irrigated area of 255 million hectares (from FAO, op. cit. note 1) by the average cost of new project construction ($7,700 per hectare), according to an analysis of 191 projects funded by the World Bank (William I. Jones, *The World Bank and Irrigation* (Washington, DC: World Bank, 1995)); Timothy C. Weiskel, "The Anthropology of Environmental Decline," Summary Statement for Hearings of the Committee on Environment and Public Works, U.S. Senate, Washington, DC, 14 September 1988.

Chapter 2. History Speaks

1. Clive Ponting, *A Green History of the World* (New York: Penguin Books, 1991).

2. Ibid.

3. Karl A. Wittfogel, *Oriental Despotism: A Comparative Study of Total Power* (New Haven, CT: Yale University Press, 1957).

4. Historians have long credited the Sumerians with the invention of writing, but recent archeological findings in Egypt suggest that Egyptian inscriptions may predate those of the Sumerians, although the evidence is not yet conclusive; see "Inscriptions Suggest Egyptians Could Have Been First to Write," *New York Times*, 16 December 1998. In either case, writing emerged from an early irrigation-based civilization.

5. Daniel Hillel, *Rivers of Eden* (New York: Oxford University Press, 1994); Ponting, op. cit. note 1.

6. H. Weiss et al., "The Genesis and Collapse of Third Millennium North Mesopotamian Civilization," *Science*, 20 August 1993.

7. Ibid.

8. Weiss et al., op. cit. note 6; John Noble Wilford, "Collapse of Earliest Known Empire is Linked to Long, Harsh, Drought," *New York Times*, 24 August 1993.

9. Height of wall from Wilford, op. cit. note 8.

10. Weiss et al., op. cit. note 6; Ann Gibbons, "How the Akkadian Empire was Hung Out to Dry," *Science*, 20 August 1993; Wilford, op. cit. note 8.

11. Thorkild Jacobsen, "Summary of Report by the Diyala Basin Archaeological Project," *Sumer*, vol. 14, pp. 79–89 (1958).

12. M.S. Drower, "Water-Supply, Irrigation, and Agriculture," in C. Singer, E.J. Holmyard, and A. R. Hall, eds., *A History of Technology* (New York: Oxford University Press, 1954).

13. Thorkild Jacobsen and Robert M. Adams, "Salt and Silt in Ancient Mesopotamian Agriculture," *Science*, 21 November 1958.
14. Ibid.
15. As quoted in Ponting, op. cit. note 1.
16. Ibid.
17. As told in Daniel J. Hillel, *Out of the Earth: Civilization and the Life of the Soil* (New York: The Free Press, 1991).
18. As quoted in Will Durant, *Our Oriental Heritage* (New York: Simon and Schuster, 1954).
19. Drower, op. cit. note 12.
20. Quote appears in K. K. Framji, B.C. Garg, and S.D.L. Luthra, *Irrigation and Drainage in the World*, Vol. 1, 3rd ed. (New Delhi: International Commission on Irrigation and Drainage, 1981); Drower, op. cit. note 12.
21. Drower, op. cit. note 12.
22. Ibid.
23. Ibid.
24. Ibid.
25. Sassanian rule from Robert M. Adams, "Agriculture and Urban Life in Early Southwestern Iran," *Science*, 13 April 1962; Sassanian irrigation and population estimate from Peter Christensen, *The Decline of Iranshahr* (Copenhagen, Denmark: Museum Tusculanum Press, University of Copenhagen, 1993); Iraq's current irrigated area from U.N. Food and Agriculture Organization, *1996 Production Yearbook* (Rome: 1997). Historian Clive Ponting (op. cit. note 1) suggests that the population may have been closer to 1.5 million, but does not explain the derivation of this estimate.
26. Sediment load from Hillel, op. cit. note 17; Christensen, op. cit. note 25.
27. Christensen, op. cit. note 25.
28. Ibid.
29. Ibid.
30. Ibid.
31. Rice as staple from Adams, op. cit. note 25; salinization episode from Jacobsen and Adams, op. cit. note 13.
32. Christensen, op. cit. note 25.
33. Framji, Garg, and Luthra, op. cit. note 20; author's visit to China and discussions with Cheng Guangwei and Zhang Junzuo, Chinese Academy of Sciences, June 1988.
34. Ponting, op. cit. note 1.
35. Hillel, op. cit. note 17.
36. Lyman P. Van Slyke, *Yangtze: Nature, History, and the River* (Menlo Park, CA: Addison-Wesley Publishing Company, 1988).

37. Ponting, op. cit. note 1.
38. Ibid.; Hillel, op. cit. note 17.
39. Malcolm Newson, *Land, Water and Development: River Basin Systems and Their Sustainable Management* (London: Routledge, 1992).
40. Early 1800s population from ibid.; current population from Population Reference Bureau, *1998 World Population Data Sheet*, wallchart (Washington, DC: 1998).
41. Karl W. Butzer, *Early Hydraulic Civilization in Egypt: A Study in Cultural Ecology* (Chicago: The University of Chicago Press, 1976); quote from Drower, op. cit. note 12.
42. Drower, op. cit. note 12.
43. Butzer, op. cit. note 41.
44. Ibid.
45. Ibid.
46. Ibid.
47. Gen. 41:1-37 (Revised Standard Version of the Bible); J. Donald Hughes, "Sustainable Agriculture in Ancient Egypt," *Agricultural History*, vol. 66, pp. 12–22 (1992); Butzer, op. cit. note 41.
48. Drower, op. cit. note 12.
49. Fekri A. Hassan, "The Dynamics of a Riverine Civilization: A Geoarchaeological Perspective on the Nile Valley, Egypt," *World Archaeology*, vol. 29, no. 1 (1997).
50. Selected portion of the hymn as quoted in Hughes, op. cit. note 47.
51. Hassan, op. cit. note 49.
52. Tax figures from Durant, op. cit. note 18; Hassan, op. cit. note 49.
53. William E. Dolittle, *Canal Irrigation in Prehistoric Mexico* (Austin: University of Texas Press, 1990).
54. Paul Kosok, *Life, Land and Water in Ancient Peru* (New York: Long Island University Press, 1965).
55. George J. Gumerman and Thomas R. Lincoln, "Preface," in George J. Gumerman, ed., *Exploring the Hohokam: Prehistoric Desert Peoples of the American Southwest* (Albuquerque: University of New Mexico Press, 1991).
56. Canal network from Paul R. Fish and Suzanne K. Fish, "Hohokam Political and Social Organization," in Gumerman, op. cit. note 55; Robert E. Gasser and Scott M. Kwiatkowski, "Recognizing Patterns in Hohokam Subsistence," in ibid.
57. Linda S. Cordell, *Prehistory of the Southwest* (Orlando, FL: Academic Press, 1984); David E. Doyel, "Hohokam Cultural Evolution in the Phoenix Basin," in Gumerman, op. cit. note 55.
58. Doyel, op. cit. note 57; Fish and Fish, op. cit. note 56; territory from Paul

R. Fish, "The Hohokam: 1000 Years of Prehistory in the Sonoran Desert." in Linda Cordell and George J. Gumerman, eds., *Dynamics of Southwest Prehistory* (Washington, DC: Smithsonian Institution Press, 1989).

59. Fish, op. cit. note 58; Doyel, op. cit. note 57.

60. Fish, op. cit. note 58; Snaketown canal from Fish and Fish, op. cit. note 56.

Chapter 3. Irrigation's Modern Era

1. Malcolm Newson, *Land, Water and Development: River Basin Systems and Their Sustainable Management* (London: Routledge, 1992).

2. World irrigation in 1800 from K.K. Framji and I.K. Mahajan, *Irrigation and Drainage in the World* (New Delhi: Caxton Press Private Ltd., 1969); 1900 and 1950 figures from William Field, "World Irrigation," *Irrigation and Drainage Systems*, vol. 4, pp. 91–107 (1990); 1995 figure from U.N. Food and Agriculture Organization (FAO), *1996 Production Yearbook* (Rome: 1997).

3. Estimate of 40 percent is approximate, and is based on a 36-percent estimate in W. Robert Rangeley, "Irrigation and Drainage in the World," in Wayne R. Jordan, ed., *Water and Water Policy in World Food Supplies* (College Station, TX: Texas A&M University Press, 1987), on a 47-percent estimate (just for grain) in Montague Yudelman, "The Future Role of Irrigation in Meeting the World's Food Supply," in Soil Science Society of America, *Soil and Water Science: Key to Understanding Our Global Environment* (Madison, WI: 1994), and on a general statement that 40 percent of world's food supply comes from irrigated land in Ismail Serageldin, *Toward Sustainable Management of Water Resources* (Washington, DC: World Bank, 1995).

4. Ian Stone, *Canal Irrigation in British India* (Cambridge, U.K.: Cambridge University Press, 1984).

5. Ibid.

6. Herbert Addison, *Land, Water, and Food* (London: Chapman & Hall, 1955); Indus flow from Frits van der Leeden, Fred L. Troise, and David Keith Todd, *The Water Encyclopedia* (Chelsea, MI: Lewis Publishers, 1990).

7. Addison, op. cit. note 6.

8. Ibid.

9. World Bank, *India Irrigation Sector Review*, Vol. 2 (Washington, DC: 1991).

10. Division of irrigated land between India and Pakistan from ibid.

11. Donald Worster, *Rivers of Empire: Water, Aridity, and the Growth of the American West* (New York: Oxford University Press, 1985).

12. Ibid.

13. Ibid.

14. Quotes and context from Stanley Roland Davison, *The Leadership of the Reclamation Movement, 1875–1902* (New York: Arno Press, 1979).

15. Charles F. Wilkinson, *Crossing the Next Meridian: Land, Water, and the Future of the West* (Washington, DC: Island Press, 1992).

16. Quotes and context from Davison, op. cit. note 14.

17. Ibid.

18. Robert M. Morgan, *Water and the Land: A History of American Irrigation* (Fairfax, VA: The Irrigation Association, 1993).

19. Quoted in Worster, op. cit. note 11.

20. U.S. Department of the Interior, Bureau of Reclamation, *50 Years of Reclamation: Financial Report to the Nation's Stockholders—1902–1952* (Washington, DC: 1952).

21. Ibid.

22. Bruce Babbitt, "The Public Interest in Western Water," *Environmental Law*, vol. 23, pp. 933–941 (1993).

23. Population in ancient Egypt from Karl W. Butzer, *Early Hydraulic Civilization in Egypt: A Study in Cultural Ecology* (Chicago: The University of Chicago Press, 1976); estimates by British demographer T. H. Hollingsworth, as cited in Joel E. Cohen, *How Many People Can the Earth Support?* (New York: W.W. Norton & Company, 1995), differ greatly and show a peak of some 30 million around A.D. 540. However, it is difficult to see how the agricultural base could have supported this large a population.

24. Daniel Hillel, *Rivers of Eden: The Struggle for Water and the Quest for Peace in the Middle East* (New York: Oxford University Press, 1994).

25. E.H. Carrier, *The Thirsty Earth: A Study in Irrigation* (London: Christophers, 1928).

26. Population in 1966 from Cohen, op. cit. note 23; Nasser quoted in Newson, op. cit. note 1.

27. High Dam height and reservoir capacity from Peter H. Gleick, ed., *Water in Crisis* (New York: Oxford University Press, 1993); Khayyám quoted in Patrick McCully, *Silenced Rivers: The Ecology and Politics of Large Dams* (London: Zed Books, 1996).

28. Current irrigated area from FAO, op. cit. note 2; addition over 20 years from "Egypt Opens New Canal to Carry Nile Water Beneath Suez to Irrigate Sinai Peninsula," *International Environmental Reporter*, 29 October 1997; current population from Population Reference Bureau (PRB),

1998 World Population Data Sheet, wallchart (Washington, DC: 1998); projected population from United Nations, *World Population Prospects 1950–2050—The 1998 Revision* (New York: December 1998).

29. Population from PRB, op. cit. note 28; water supplies from World Resources Institute, *World Resources Report 1994–95* (New York: Oxford University Press, 1994).

30. Zhang Zezhen and Deng Shangshi, "The Development of Irrigation in China," *Water International,* June 1987.

31. Ministry of Water Resources and Electric Power, People's Republic of China, *Irrigation and Drainage in China* (Beijing: China Water Resources and Electric Power Press, 1987); role of peasants from Vaclav Smil, "China's Water Resources," *Current History*, September 1979.

32. Mao quoted in Smil, op. cit. note 31; People's Victory Irrigation District from Ministry of Water Resources and Electric Power, op. cit. note 31; James Nickum, "Issue Paper on Water and Irrigation," prepared for "The Strategy and Action Project for Chinese and Global Food Security," Millennium Institute, Arlington, VA, 1997.

33. Ministry of Water Resources and Electric Power, op. cit. note 31; FAO, op. cit. note 2.

34. World Bank, op. cit. note 9; Ruth Meinzen-Dick, *Groundwater Markets in Pakistan: Participation and Productivity*, Research Report 105 (Washington DC: International Food Policy Research Institute (IFPRI), 1996).

35. World Bank, op. cit. note 9; FAO, op. cit. note 2; Indus Basin area from Meinzen-Dick, op. cit. note 34.

36. World Bank, "Irrigation in the USSR," unpublished manuscript, September 1990.

37. Ibid.; cotton trends from Zhores A. Medvedev, *Soviet Agriculture* (New York: W.W. Norton & Company, 1987).

38. World Bank, op. cit. note 36; FAO, op. cit. note 2; U.S. Department of Agriculture (USDA), Economic Research Service (ERS), *USSR: Agriculture and Trade Report* (Washington, DC: 1989).

39. Figure for 1950 from USDA, ERS, "Irrigation, Nation's Largest Water Use," *Agricultural Resources Situation & Outlook* (Washington, DC: 1993); 1995 figure from FAO, op. cit. note 2; Bureau of Reclamation share from Richard W. Wahl, *Markets for Federal Water: Subsidies, Property Rights, and the Bureau of Reclamation* (Washington DC: Resources for the Future, 1989); Dominy quote from "Last Oasis," fourth episode of the *Cadillac Desert* series, Public Broadcasting System, aired July 1997.

40. National Research Council (NRC), *A New Era for Irrigation* (Washington, DC: National Academy Press, 1996); Bureau figures from USDA, op. cit. note 39.

41. NRC, op. cit. note 40; shares of cropland irrigated from FAO, op. cit. note 2.

42. Growth rates in recent decades from Mark Rosegrant, *Water Resources in the Twenty-First Century: Challenges and Implications for Action* (Washington DC: IFPRI, 1997); 17-percent drop by 2020 assumes global irrigated area expands by some 40 million hectares, as projected in Per Pinstrup-Anderson, Rajul Pandya-Lorch, and Mark W. Rosegrant, *The World Food Situation: Recent Developments, Emerging Issues, and Long-Term Prospects* (Washington, DC: IFPRI, 1997); 28-percent drop by 2020 assumes no net growth in irrigated area by 2020—that is, gains are offset by losses; irrigation trends from FAO, *Production Yearbook*, various years; population from United Nations, op. cit. note 28.

43. Mark Rosegrant, Claudia Ringler, and Roberta V. Gerpacio, "Water and Land Resources and Global Food Supply," paper prepared for the 23rd International Conference of Agricultural Economists on Food Security, Diversification, and Resource Management: Refocusing the Role of Agriculture? Sacramento, CA, 10–16 August 1997.

44. William I. Jones, *The World Bank and Irrigation* (Washington, DC: World Bank, 1995).

45. African country areas from FAO, op. cit. note 2.

46. Donor investment trends from Rosegrant, op. cit. note 42.

47. Share needing repair from World Bank, "Information Brief on Irrigation and Drainage" (Washington, DC: 1993); cost comparison from Jones, op. cit. note 44.

48. Daniel P. Beard, *Blueprint for Reform: The Commissioner's Plan for Reinventing Reclamation* (Washington, DC: Bureau of Reclamation, 1993); U.S. Department of the Interior, "Bureau of Reclamation Announces Reforms: Meets Challenge of the National Performance Review," press release, (Washington, DC: 1 November 1993).

Chapter 4. Running Out

1. "Yellow River Weeps for Water," *China Daily*, 10 February 1998; flow stoppage count from "Huang He River Flow Stops in Lower Reaches," *Xinhua Press* (via Foreign Broadcast Information Service (FBIS), <http://fbis.fedworld.gov/cgi-bin/retrieve>, 5 September 1997; 226 days from Robert J. Saiget, "China Congress Tackling Water Shortages, Pollution," *Kyoto News* (via FBIS), 11 March 1998.

2. Han Zhenjun and Meng Min, "Is the Huang He Becoming a Seasonal River?" Part 1 of four-part series, "China: Survey on Huang He Drying Up," *Jingji Cankao Bao* (via FBIS), 15 August 1997; "Yellow River Weeps

for Water," op. cit. note 1.

3. Robert M. Morgan, *Water and the Land: A History of American Irrigation* (Fairfax, VA: The Irrigation Association, 1993).

4. Han and Meng, op. cit. note 2; irrigated area from "China's Second Largest River Threatened by Ongoing Industrial, Agricultural Diversion," *International Environment Reporter*, 18 February 1998.

5. Demand projections from Dennis Engi, Sandia National Laboratories, Albuquerque, NM, discussion with Brian Halweil, Worldwatch Institute, April 1998; Shandong Province from Fred Crook, Economic Research Service, U.S. Department of Agriculture (USDA), Washington, DC, discussion with Brian Halweil, Worldwatch Institute, April 1998.

6. "Yellow River Weeps for Water," op. cit. note 1; 1997 losses from Saiget, op. cit. note 1; quote from Liu Zhenying and Wang Yanbing, "China: Jiang Chunyun on Huang He Water Issues," *Xinhua Domestic Service* (via FBIS), 2 October 1997.

7. Author's visit to Yellow River Conservancy Commission, Zhengzhou, China, 8 June 1988; Ministry of Water Resources and Electric Power, People's Republic of China, *Irrigation and Drainage in China* (Beijing: China Water Resources and Electric Power press, 1987).

8. Patrick McCully, *Silenced Rivers: The Ecology and Politics of Large Dams* (London: Zed Books, 1996).

9. Zhu Qiwen, "Water-Control Dream Fulfilled," *China Daily*, 22 December 1997; Liu Bin and Zhao Jiang, "China: Feature on Huang He, Chang Jiang," *Xinhua Press* (via FBIS), 28 October 1997; "China: Xiaolangdi Dam to Mitigate Huang He Flood Damage," *Xinhua Press* (via FBIS), 15 August 1997; cost and resettlement figures from "China's Xiaolangdi Project Caps Centuries of Flood Control Efforts on Yellow River," *World Bank News*, 21 April 1994.

10. Expected siltation from James Nickum, "Issue Paper on Water and Irrigation," prepared for The Strategy and Action Project for Chinese and Global Food Security, Millennium Institute, Arlington, VA, 1997.

11. For background, see Asit K. Biswas et al., eds., *Long-Distance Water Transfer: A Chinese Case Study and International Experience* (Dublin, Ireland: Tycooly International, 1983); Sandra Postel, *Water: Rethinking Management in an Age of Scarcity*, Worldwatch Paper 62 (Washington, DC: Worldwatch Institute, December 1984); current status from Brian Halweil, Worldwatch Institute, discussion with author, April 1998.

12. Lower-end volume from Wang and Shen, op. cit. note 5; upper end and cost projections from Engi, op. cit. note 5.

13. Patrick Tyler, "China's Fickle Rivers: Dry Farms, Needy Industry Bring a Water Crisis," *New York Times*, 23 May 1996.

14. Sandra Postel, "Where Have All the Rivers Gone?" *World Watch*, May-June 1995; Harald Frederiksen, Jeremy Berkoff, and William Barber, *Water Resources Management in Asia* (Washington, DC: World Bank, 1993).

15. Figure of 1.3 billion cubic meters (bcm) is best estimate based on 3.5 bcm freshwater outflow in 1987 and projected 0.6 bcm outflow projected for 2000, from R. Stoner, "Future Irrigation Planning in Egypt," in P.P. Howell and J.A. Allan, eds., *The Nile: Sharing a Scarce Resource* (Cambridge, U.K.: Cambridge University Press, 1994); Postel, op. cit. note 14.

16. Mark Huband, "Egypt a Step Nearer to Taming the Nile," *Financial Times*, 20 February 1998; "Egypt Opens New Canal to Carry Nile Water Beneath Suez to Irrigate Sinai Peninsula," *International Environment Reporter*, 29 October 1997.

17. Fred Pearce, "High and Dry in Aswan," *New Scientist*, 7 May 1994.

18. Robin M. Leichenko and James L. Wescoat, Jr., "Environmental Impacts of Climate Change and Water Development in the Indus Delta Region," *Water Resources Development*, vol. 9, no. 3 (1993).

19. Frederiksen, Berkoff, and Barber, op. cit. note 14; threats to Bengal tiger from M. Roushanuzzaman, "Water Straining Relations Between India-Bangladesh," *Depthnews Asia* (Manila), June 1995.

20. Frederiksen, Berkoff, and Barber, op. cit. note 14.

21. National Environmental Engineering Research Institute (NEERI),"Water Resources Management in India: Present Status and Solution Paradigm," Nagpur, India, undated (circa 1997).

22. "Alarming Ground Water Depletion in Haryana and Punjab," *IARI News* (Indian Agricultural Research Institute), October-December 1993; Coimbatore and Mehsana district depletion from Tata Energy Research Institute, *Looking Back to Think Ahead: Executive Summary* (New Delhi: undated); Sandra Postel, *Last Oasis*, rev. ed. (New York: W.W. Norton & Company, 1997).

23. NEERI, op. cit. note 21; David Seckler, David Molden, and Randolph Barker, *Water Scarcity in the Twenty-First Century*, Water Brief No. 1 (Colombo, Sri Lanka: International Water Management Institute, 1998).

24. "Alarming Ground Water Depletion in Haryana and Punjab," op. cit. note 22; Surendar Singh, "Some Aspects of Groundwater Balance in Punjab," *Economic and Political Weekly*, 28 December 1991.

25. L.R. Khan, "Environmental Impacts of Groundwater Development in Bangladesh," paper presented at the IX World Water Congress of the International Water Resources Association, Montreal, Canada, 1–6 September 1997; Rural Development Sector Unit, South Asia Region, *Water*

Resource Management in Bangladesh: Steps Toward a New National Water Plan (Dhaka, Bangladesh: World Bank, 1998); excess suffering of poor farmers from author's participation in seminar of Bangladeshi water experts, organized by International Development Enterprises, Dhaka, Bangladesh, 16 January 1998.

26. Zhang Qishun and Zhang Xiao, "Water Issues and Sustainable Social Development in China," *Water International*, vol. 20, no. 3 (1995); Liu Yonggong and John B. Penson, "China's Sustainable Agriculture and Regional Implications," paper presented to the Symposium on Agriculture, Trade and Sustainable Development in Pacific Asia: China and its Trading Partners (no place or date given); Sandia National Laboratories, China Infrastructure Initiative: Decision Support Systems, <http://www.igaia.sandia.gov/ igaia/china/chinamodel.html>.

27. River basin demand and supply data and projections from Engi, op. cit. note 5.

28. USDA, *Production Supply, and Distribution*, electronic database, Washington, DC, updated February 1999.

29. Ruth Meinzen-Dick, *Groundwater Markets in Pakistan: Participation and Productivity* (Washington, DC: International Food Policy Research Institute, 1996); total and annual net depletion based on data in Edwin D. Gutentag et al., *Geohydrology of the High Plains Aquifer in Parts of Colorado, Kansas, Nebraska, New Mexico, Oklahoma, South Dakota, Texas, and Wyoming* (Washington, DC: U.S. Government Printing Office, 1984), and in Dork L. Sahagian, Frank W. Schwartz, and David K. Jacobs, "Direct Anthropogenic Contributions to Sea Level Rise in the Twentieth Century," *Nature*, 6 January 1994.

30. Irrigated area figures from National Research Council, *A New Era for Irrigation* (Washington DC: National Academy Press, 1996).

31. California Department of Water Resources, *California Water Plan Update*, vol 1 (Sacramento, CA: 1994).

32. Postel, op. cit. note 22; water demand from A.S. Al-Turbak, "Meeting Future Water Shortages in Saudi Arabia," paper presented at the IX World Water Congress of the International Water Resources Association, Montreal, Canada, 1–6 September 1997; Charles J. Hanley, "Saudi Arabia Farming Sucks the Country Dry," *Associated Press*, 29 March 1997; peak grain production from USDA, op. cit. note. 28.

33. USDA, op. cit. note 28.

34. Peak water deficit from Al-Turbak, op. cit. note 32; likely current deficit is author's estimate based on mid- to late 1980s figures in Abdulla Ali Al-Ibrahim, "Excessive Use of Groundwater Resources in Saudi Arabia: Impacts and Policy Options," *Ambio*, February 1991.

35. African Sahara depletion from Sahagian, Schwartz, and Jacobs, op. cit. note 29; Libya's depletion calculated from Rajab M. El Asswad, "Agricultural Prospects and Water Resources in Libya," *Ambio*, September 1995; "GMR Wins Prestigious Pipeline Award," *World Water and Environmental Engineering*, July 1997.

36. "Libyan GMR Military Claims Ridiculed," *World Water and Environmental Engineering*, February 1998; Postel, op. cit. note 22; 80-percent figure from Roula Khalaf, "Gadaffi Taps Desert Waters in Bid to Make a Big Splash," *Financial Times*, 11 September 1996; Fred Pearce, "Will Gaddafi's Great River Run Dry?" *New Scientist*, September 1991.

37. Quotes from Pearce, op. cit. note 36; Raymond Bonner, "Libya's Vast Desert Pipeline Could be Conduit for Troops," *New York Times*, 2 December 1997; "Libyan GMR Military Claims Ridiculed," op. cit. note 36.

38. Inaccessible runoff from Sandra L. Postel, Gretchen C. Daily, and Paul R. Ehrlich, "Human Appropriation of Renewable Fresh Water," *Science*, 9 February 1996.

39. Figure of 70 percent from ibid.; number of dams from Engelbertus Oud and Terence C. Muir, "Engineering and Economic Aspects of Planning, Design, Construction and Operation of Large Dam Projects," in Tony Dorcey, ed., *Large Dams: Learning from the Past, Looking at the Future* (Gland, Switzerland, and Washington, DC: World Conservation Union–IUCN and World Bank, 1997); 6,600 cubic kilometer figure is author's estimate, based on an estimate of 5,500 cubic kilometers from M.I. L'Vovich et al., in B.L. Turner et al., *The Earth as Transformed by Human Action* (Cambridge, U.K.: Cambridge University Press, 1990) and an assumption of a 20-percent increase since this estimate was made.

40. McCully, op. cit. note 8.

41. Number of dams from Omar Sattaur, "India's Troubled Waters," *New Scientist*, 27 May 1989.

42. J. Patel, "Who Benefits Most from Damming the Narmada," *Economic and Political Weekly* (India), 29 December 1990; McCully, op. cit. note 8.

43. Japan pullout from McCully, op. cit. note 8.

44. "World Bank to Assess Narmada," *World Rivers Review*, March/April 1991; Bradford Morse and Thomas R. Berger, *Sardar Sarovar*, Report of the Independent Review (Ottawa, ON: Resource Futures International, Inc., 1992); World Bank pullout from McCully, op. cit. note 8.

45. McCully, op. cit. note 8.

46. Ibid; interim order from Kalpana Sharma, "Too Early to Rejoice over Narmada Verdict?" *The Hindu*, 24 February 1999.

47. A thoughtful expression of this perspective is that of David Seckler, "The Sardar Sarovar Project in India: A Commentary on the Report of the Independent Review," Discussion Paper No. 8, Center for Economic Policy Studies, Winrock International Institute for Agricultural Development, Arlington, VA, July 1992.

48. Ibid.

49. See Thayer Scudder, "Social Impacts of Large Dam Projects," in Dorcey, op. cit. note 39, and McCully, op. cit. note 8; Harald D. Frederiksen, "International Community Response to Critical World Water Problems: A Perspective for Policy Makers," *Water Policy*, vol. 1, pp. 139–58 (1998).

50. McCully, op. cit. note 8.

51. Dorcey, op. cit. note 39.

52. Nizamsagar from Malcolm Newson, *Land, Water and Development: River Basin Systems and Their Sustainable Management* (London: Routledge, 1992); Tarbela from McCully, op. cit. note 8.

53. K. Mahmood, *Reservoir Sedimentation: Impact, Extent, and Mitigation* (Washington, DC: World Bank, 1987); replacement cost assumes 15–20¢ per cubic meter; dredging costs of $2–3 per cubic meter from ibid.

54. For general background, see Stephen H. Schneider, *Global Warming: Are We Entering the Greenhouse Century?* (San Francisco, CA: Sierra Club Books, 1989), and Paul E. Waggoner, ed., *Climate Change and U.S. Water Resources* (New York: John Wiley & Sons, 1990); Thomas R. Karl, Neville Nicholls, and Jonathan Gregory, "The Coming Climate," *Scientific American,* May 1997.

55. Postel, op. cit. note 22.

56. For an overview, see P. H. Gleick, "Climate Change, Hydrology, and Water Resources," *Review of Geophysics*, vol. 27, no. 3 (1989), and Leichenko and Wescoat, op. cit. note 18.

57. John Schaake, "From Climate to Flow," in Waggoner, op. cit. note 54.

58. Figure of $200–400 billion assumes cost range of 15–30¢ per cubic meter of new capacity; Karl, Nicholls, and Gregory, op. cit. note 54; estimate of additional $120 billion assumes irrigation project costs average $4,800 per hectare, from William I. Jones, *The World Bank and Irrigation* (Washington, DC: World Bank, 1995).

59. Peter H. Gleick, "Vulnerability of Water Systems," in Waggoner, op. cit. note 54.

60. Effect on stomata from H.A. Mooney et al., "Predicting Ecosystem Responses to Elevated CO_2 Concentrations," *BioScience*, February 1991; effect of lack of soil moisture from Cynthia Rosenzweig and Daniel Hillel, "Agriculture in a Greenhouse World," *Research & Exploration*, vol. 9, no. 2 (1993).

61. Cynthia Rosenzweig and Martin L. Parry, "Potential Impact of Climate Change on World Food Supply," *Nature*, 13 January 1994.

Chapter 5. A Faustian Bargain

1. Christopher Marlowe, *Dr. Faustas* (ca. 1592–93), in M.H. Abrams et al., eds., *The Norton Anthology of English Literature* (New York: W.W. Norton & Company, 1986).

2. Figure of $11 billion from F. Ghassemi, A.J. Jakeman, and H.A. Nix, *Salinisation of Land and Water Resources: Human Causes, Extent, Management and Case Studies* (Sydney: University of New South Wales Press Ltd, 1995); 2 million hectares a year from Dina L. Umali, *Irrigation-Induced Salinity* (Washington, DC: World Bank, 1993).

3. Author's visit to the region as a member of the International Advisory Panel for the Aral Sea Program of the World Bank, March 1995.

4. For fuller description, see Sandra Postel, *Dividing the Waters: Food Security, Ecosystem Health, and the New Politics of Scarcity*, Worldwatch Paper 132 (Washington, DC: Worldwatch Institute, September 1996); U.N. Development Programme (UNDP), *Aral in Crisis* (Tashkent: 1995).

5. Figure of 120 million from Don Blackmore, Murray-Darling River Basin Commission, and Member, International Advisory Panel, Aral Sea Program of the World Bank, panel meeting, Washington, DC, 20 December 1996; salinity in Syr Darya from Ghassemi, Jakeman and Nix, op. cit. note 2.

6. Aral Sea Basin Program—Group 1, Interstate Commission for Water Coordination, and World Bank, "Developing a Regional Water Management Strategy: Issues and Work Plan," draft prepared for the Executive Committee of the Interstate Council for the Aral Sea, May 1996.

7. Ibid.; Karakalpakstan yields from UNDP, op. cit. note 4.

8. Turkmenistan from Aral Sea Basin Program, Interstate Commission, and World Bank, op. cit. note 6; figure of 2 million hectares from discussion among members of International Advisory Panel for the Aral Sea Program of the World Bank, panel meeting, Washington, DC, 20 December 1996.

9. Early geologic history from Ghassemi, Jakeman, and Nix, op. cit. note 2.

10. Situation in 1947 from Asit K. Biswas, "Environmental Concerns in Pakistan, with Special Reference to Water and Forests," *Environmental Conservation*, winter 1987; plan in 1961 from Ghassemi, Jakeman and Nix, op. cit. note 2.

11. Biswas, op. cit. note 10.

12. Ibid.

13. Figure of 43 percent from ibid.; 60 percent from Ruth Meinzen-Dick,

Groundwater Markets in Pakistan: Participation and Productivity (Washington, DC: International Food Policy Research Institute, 1996); $1 billion figure from Umali, op. cit. note 2.

14. Meinzen-Dick, op. cit. note 13.

15. Jacob W. Kijne,*Water and Salinity Balances for Irrigated Agriculture in Pakistan* (Colombo, Sri Lanka: International Water Management Institute, 1996).

16. Ibid.

17. Population growth from Population Reference Bureau, *1998 World Population Data Sheet*, wallchart (Washington, DC: 1998); Kijne, op. cit. note 15.

18. H. E. Dregne, Zhixun Xiong, and Siyu Xiong, "Soil Salinity in China," *Desertification Control Bulletin*, no. 28, 1996.

19. Ghassemi, Jakeman, and Nix, op. cit. note 2.

20. Richard W. Wahl, *Markets for Federal Water: Subsidies, Property Rights, and the Bureau of Reclamation* (Washington, DC: Resources for the Future, 1989).

21. Estimated capital cost of plant from ibid.; Ghassemi, Jakeman, and Nix, op. cit. note 2.

22. National Research Council (NRC), *A New Era for Irrigation* (Washington, DC: National Academy Press, 1996).

23. Harrison C. Dunning, "Confronting the Environmental Legacy of Irrigated Agriculture in the West: The Case of the Central Valley Project," *Environmental Law*, vol. 23, pp. 943–69 (1993); Tom Harris, in *Death in the Marsh* (Washington, DC: Island Press, 1991), provides a detailed and gripping account.

24. Elliot Marshall, "High Selenium Levels Confirmed in Six States," *Science*, 10 January 1986; Ted Williams, "Death in a Black Desert," *Audubon*, January-February 1994.

25. Williams, op. cit. note 24; California Department of Water Resources, *California Water Plan Update*, Vol. 1, Bulletin 160–93 (Sacramento, CA: 1994).

26. William I. Jones, *The World Bank and Irrigation* (Washington, DC: World Bank, 1995).

27. Ghassemi, Jakeman, and Nix, op. cit. note 2.

28. Robert H. Boyle, "Life—or Death—for the Salton Sea?" *Smithsonian*, June 1996.

29. Ibid.; Verne G. Kopytoff, "In Spring, Birds Return to the Salton Sea and Die in Droves," *New York Times*, 24 March 1998.

30. Ghassemi, Jakeman, and Nix, op. cit. note 2; San Joaquin valley pond area from M.E. Grismer, F. Karajeh, and H. Bower, "Evaporation Pond

Hydrology," in Richard G. Allen, ed., *Management of Irrigation and Drainage Systems: Integrated Perspectives* (New York: American Society of Civil Engineers, 1993).

31. In dry regions, higher irrigation efficiency can increase salt problems if not accompanied by adequate leaching of salts from the soil; see A. Banin and A. Fish, "Secondary Desertification due to Salinization of Intensively Irrigated Lands: The Israeli Experience," *Environmental Monitoring and Assessment*, vol. 37, pp. 17–37 (1995).

32. Israeli examples from A. Benin, "Utilization of Recycled, Saline and Other Marginal Waters for Irrigation: Challenges and Management Issues," in F. Lopez-Vera, J. De Castro Morcillo, and A. Lopez Lillio, eds., *Uso del Agua en las Areas Verdes Urbanas* (Madrid: 1993).

33. J.D. Rhoades et al., "Use of Saline Drainage Water for Irrigation: Imperial Valley Study," *Agricultural Water Management*, vol. 16, pp. 25–36 (1989).

34. Ibid.

35. Ghassemi, Jakeman, and Nix, op. cit. note 2.

36. A.F. Heuperman and V. Cervinka, "Integrated Biological and Engineering Systems for Salinity Management in Australia and California," unpublished paper, August 1992; K.K. Tanji and F.F. Karajeh, "Saline Drain Water Reuse in Agroforestry Systems," *Journal of Irrigation and Drainage Engineering*, January-February 1993.

37. NRC, op. cit. note 22.

38. NRC, *Saline Agriculture: Salt-Tolerant Plants for Developing Countries* (Washington, DC: National Academy Press, 1990); Edward P. Glenn, J. Jed Brown, and James W. O'Leary, "Irrigating Crops with Seawater," *Scientific American*, August 1998.

39. Ghassemi, Jakeman, and Nix, op. cit. note 2.

40. For more on salinity assessments, see J.D. Rhoades, "Sustainability of Irrigation: An Overview of Salinity Problems and Control Strategies," presented at Footprints of Humanity: Annual Conference of the Canadian Water Resources Association, Lethbridge, AB, Canada, 3–6 June 1997, and J.D. Rhoades, "Measuring and Monitoring Soil Salinity," in A. Kandiah, ed., *Water, Soil and Crop Management Relating to the Use of Saline Water* (Rome: U.N. Food and Agriculture Organization, 1990).

41. United Nations, *World Population Prospects 1950–2050: The 1998 Revision* (New York: December 1998).

Chapter 6. Water Wars I: Farms Versus Cities and Nature

1. Charles McCoy and G. Pascal Zachary, "A Bass Play in Water May Presage Big Shift in its Distribution," *Wall Street Journal*, 11 July 1997.

2. Ibid.; Imperial Irrigation District (IID) water cost from author's visit to IID, Imperial County, CA, 2 May 1996; Peter Passell, "A Gush of Profits from Water Sale?" *New York Times*, 23 April 1998.

3. Jason Morrison, Sandra Postel, and Peter Gleick, *The Sustainable Use of Water in the Lower Colorado River Basin* (Oakland, CA: Pacific Institute, 1996).

4. Sale in 1997 from Passell, op. cit. note 2, and from "Valley of the Dammed," *The Economist*, 21 February 1998; "Water in California: Flowing Gold," *The Economist*, 10 October 1998; initial price of transferred water from Sue McClurg, "Cutting Colorado River Use: The California Plan," *Western Water*, November-December 1998.

5. Twain quote from Barbara K. Rodes and Rice Odell, *A Dictionary of Environmental Quotations* (New York: Simon & Schuster, 1992).

6. United Nations, Population Division, *World Urbanization Prospects: The 1996 Revision* (New York: 1998).

7. Ibid.; Mark W. Rosegrant and Claudia Ringler, "Impact on Food Security and Rural Development of Reallocating Water from Agriculture for Other Uses," paper prepared for Harare Expert Group Meeting on Strategic Approaches to Freshwater Management, Harare, Zimbabwe, 28–31 January 1998.

8. Rosegrant and Ringler, op. cit. note 7.

9. Patrick E. Tyler, "China Lacks Water to Meet its Mighty Thirst," *New York Times*, 7 November 1993.

10. Figure for 1949 from Zhang Zezhen et al., "Challenges to and Opportunities for Development of China's Water Resources in the 21st Century," *Water International*, vol. 17, no. 1 (1992); sources differ on both the present number of China's cities (generally between 600 and 640) and the number that are water-short (generally between 300 and 480); planning study cited in James E. Nickum, "Issue Paper on Water and Irrigation," prepared for the project on Strategy and Action for Chinese and Global Food Security, Millennium Institute, U.S. Department of Agriculture (USDA), World Bank, and Worldwatch Institute, Arlington, VA, September 1997; Eugene Linden, "The Exploding Cities of the Developing World," *Foreign Affairs*, January-February 1996; United Nations, op. cit. note 6.

11. Lester R. Brown and Brian Halweil, "China's Water Shortage Could Shake World Food Security," *World Watch*, July-August 1998; Huai and Smil cites from Joseph Kahn, "China's 'Greens' Win Rare Battle on River," *Wall Street Journal*, 2 August 1996.

12. Urban increase calculated from Population Reference Bureau (PRB), *1998 Population Data Sheet*, wallchart (Washington, DC: 1998), and from United Nations, op. cit. note 6; Tirupur from Jan Lundqvist, "Food

Production and Food Security in an Urbanising World," prepared for the IXth World Water Congress of the International Water Resources Association, Montreal, Canada, 1–6 September 1997.

13. Ganjar Kurnia, Teten Avianto, and Bryon Bruns, "Farmers, Factories and the Dynamics of Water Allocation in West Java," in Bryan Bruns and Ruth S. Meinzen-Dick, eds., *Negotiating Water Rights* (New Delhi: Sage Publishers, forthcoming).

14. Sandra Postel, *Dividing the Waters: Food Security, Ecosystem Health, and the New Politics of Scarcity*, Worldwatch Paper 132 (Washington, DC: Worldwatch Institute, September 1996).

15. Santos Gomez and Anna Steding, "California Water Transfers: An Evaluation of the Economic Framework and a Spatial Analysis of the Potential Impacts," Pacific Institute, Oakland, CA, April 1998.

16. Penn Loh and Anna Steding, "The Palo Verde Test Land Fallowing Program: A Model for Future California Water Transfers?" Pacific Institute, Oakland, CA, March 1996.

17. Ibid.

18. "Clock Ticks for Dying Salton Sea; Saving It Would Make Prehistory," *U.S. Water News*, January 1999; McClurg, op. cit. note 4.

19. Rosegrant and Ringler, op. cit. note 7; R. Meinzen-Dick, "Valuing the Multiple Uses of Irrigation Water," in Melvyn Kay, Tom Franks, and Laurence Smith, eds., *Water: Economics, Management and Demand* (London: E & F N Spon, 1997).

20. Renato Gazmuri Schleyer and Mark W. Rosegrant, "Chilean Water Policy: The Role of Water Rights, Institutions, and Markets," in Mark W. Rosegrant and Renato Gazmuri Schleyer, *Tradable Water Rights: Experiences in Reforming Water Allocation Policy* (Arlington, VA: Irrigation Support Project for Asia and the Near East, 1994).

21. Mats Dynesius and Christer Nilsson, "Fragmentation and Flow Regulation of River Systems in the Northern Third of the World," *Science*, 4 November 1994.

22. Global extinction threats from Jonathan Baillie and Brian Groombridge, eds., *1996 IUCN Red List of Threatened Animals* (Gland, Switzerland: World Conservation Union–IUCN, 1996); B.A. Stein and S.R. Flack, *1997 Species Report Card: The State of U.S. Plants and Animals* (Arlington, VA: The Nature Conservancy, 1997).

23. Michael R. Moore, Aimee Mulville, and Marcia Weinberg, "Water Allocation in the American West: Endangered Fish Versus Irrigated Agriculture," *Natural Resources Journal*, spring 1996.

24. Ibid.

25. Ibid.; Charles F. Wilkinson, *Crossing the Next Meridian* (Washington,

DC: Island Press, 1992).

26. Postel, op. cit. note 14; Sue McClurg, "Saving the Salmon," *Western Water*, January-February 1998.

27. "Solemn News on Salmon," *Amicus Journal*, spring 1998; "Keeping Wetlands Wet," *Amicus Journal*, fall 1998.

28. Sue McClurg, "Delta Debate," *Western Water*, March-April 1998.

29. Ibid.; "CALFED Releases Revised Delta Plan," *Western Water*, November/December 1998.

30. "Teeming Oasis or Desert Mirage: Bringing a Wildlife Refuge Back to Life," *Nature Conservancy*, September-October 1991; Timothy Egan, "Where Water Is Power, the Balance Shifts," *New York Times*, 30 November 1997.

31. Quote from Egan, op. cit. note 30; "Novel Use of Clean-Water Loans Brightens Outlook for a River," *New York Times*, 31 October 1996.

32. "Victory for the Fishes," *The Economist*, 6 December 1997; Mary DeSena, "Edwards Dam May be Removed to Restore Fisheries on the Kennebec," *U.S. Water News*, October 1997; Paul Koberstein, "Dam Slayers Have their Day," *Inner Voice*, September-October 1997.

33. *Idaho Statesman* cited in Susan Whaley, "An Idaho Daily Breaches the Northwest's Silence over Tearing Down Dams," *High Country News*, 1 September 1997.

34. Number of ESA listings from Mary DeSena, "Feds Want ESA Listing for 13 Salmon, Steelhead Species," *U.S. Water News*, May 1998; 200 runs extinct from Marc Reisner, "Coming Undammed," *Audubon*, September-October 1998; Babbitt quote from U.S. Department of Interior, "Interior Secretary Signs Landmark Conservation Agreement to Remove Edwards Dam," press release (Washington, DC: 26 May 1998).

35. For description of Glen Canyon proposal, see David R. Brower, "Let the River Run Through It," *Sierra*, March-April 1997; Daniel P. Beard, "Dams Aren't Forever," *New York Times*, 6 October 1997.

36. Michael Collier, Robert H. Webb, and John C. Schmidt, *Dams and Rivers: Primer on the Downstream Effects of Dams*, Circular 1126 (Denver, CO: U.S. Geological Survey, 1996).

37. Ibid.

38. "New Regs Put into Effect for Managing Colorado R.," *U.S. Water News*, January 1997; William K. Stevens, "A Dam Open, Grand Canyon Roars Again," *New York Times*, 25 February 1997.

39. "Water Capping under Discussion in Australia," *World Water and Environmental Engineering*, June 1997.

40. Tim Fisher, "Fish Out of Water: The Plight of Native Fish in the Murray-Darling," *Habitat Australia*, October 1996.

41. "Water Capping under Discussion in Australia," op. cit. note 39.

42. Interstate Council for the Aral Sea (ICAS), "Developing a Regional Water Management Strategy: Issues and Work Plan" (draft), in cooperation with World Bank, Washington, DC, May 1996.

43. Figure of 35 billion from N.F. Glazovskiy, "Ideas on an Escape from the 'Aral Crisis'," *Soviet Geography*, February 1991; calculation of cropland required assumes consumptive water use of 8,000 cubic meters per hectare per year, about 65 percent of average irrigation withdrawals, which could be conservative; 13–19 billion figure from ICAS, op. cit. note 42.

44. Continued hopes for Siberian diversion from author's discussions with Nikita F. Glazovskiy, Deputy Director of the Institute of Geography of the Russian Academy of Sciences during a visit to the Aral Sea region in March 1995, and also from ICAS, op. cit. note 42; Siberian diversion from Sandra Postel, *Last Oasis*, rev. ed. (New York: W.W. Norton & Company, 1997); plans to expand irrigated area from ICAS, op. cit. note 42.

45. J.A. Allan, "The Political Economy of Water: Reasons for Optimism, But Long Term Caution," in J.A. Allan and J.H.O. Court, eds., *Water in the Jordan Catchment Countries* (London: School of Oriental and Asian Studies, University of London, 1995).

46. Sandra L. Postel, "Water for Food Production: Will There by Enough in 2025?" *BioScience*, August 1998; annual net grain imports for these countries, averaged over 1994–96, totaled 48 million tons. Table 6–1 from the following: Africa runoff from U.N. Food and Agriculture Organization (FAO), *Irrigation in Africa in Figures*, Water Report No. 7 (Rome: 1995); Mideast and Asian runoff from World Resources Institute (WRI), *World Resources 1994–95* (New York: Oxford University Press, 1994); net grain imports and consumption from USDA, *Production, Supply, and Distribution*, electronic database, Washington, DC, updated February 1999.

47. Postel, op. cit. note 46. Table 6–2 from the following: Africa runoff from FAO, op. cit. note 46; Mideast and Asian runoff from WRI, op. cit. note 46; 1995 population from PRB, *1995 World Population Datasheet*, wallchart (Washington, DC: 1995); projected 2025 population from PRB, op. cit. note 12.

48. World Food Programme, *Tackling Hunger in a World Full of Food: Tasks Ahead for Food Aid* (Rome: 1996); 1996–97 food aid from International Commission on Irrigation and Drainage, *ICID News Update*, New Delhi, India, November 1997.

Chapter 7. Water Wars II: Irrigation and the Politics of Scarcity

1. Stephen McCaffrey, "Water Scarcity: Institutional and Legal Responses," in Edward H. P. Brans et al., eds., *The Scarcity of Water: Emerging Legal and Policy Responses* (The Hague, Netherlands: Kluwer Law International, 1997); 150 years from Clive Ponting, *A Green History of the World* (New York: Penguin Books, 1991).

2. See, for example, the chronology in Peter H. Gleick, *The World's Water 1998–99* (Washington, DC: Island Press, 1998).

3. Sadat quote from Joyce R. Starr, "Water Wars," *Foreign Policy*, March 1991; Boutros-Ghali quote from Peter H. Gleick, "Water, War & Peace in the Middle East," *Environment*, April 1994; King Hussein statement from Joyce R. Starr, "Nature's Own Agenda: A War for Water in the Mideast," *Washington Post*, 3 March 1991; Serageldin quote from press release, World Bank, Washington, DC, 1995.

4. Thomas F. Homer-Dixon, "Environmental Scarcities and Violent Conflict," *International Security*, summer 1994; quote from Thomas Homer-Dixon, "The Project on Environment, Population and Security: Key Findings of Research," *Report of the Environmental Change and Security Project*, Woodrow Wilson Center, Washington, DC, spring 1996.

5. Water deficit figures from Asit K. Biswas et al., *Core and Periphery: A Comprehensive Approach to Middle Eastern Water* (Delhi: Oxford University Press, 1997).

6. Ibid.; Syria's population from Population Reference Bureau (PRB), *1998 Population Data Sheet*, wallchart (Washington, DC: 1998).

7. Biswas et al., op. cit. note 5; Jordan-Israel Peace Treaty, *Annex II, Water Related Matters*, 17 October 1994; Frederic Hof, "Jordan's Water Diplomacy," *Journal of Commerce*, 18 June 1997; BBC news report, Amman, Jordan, 23 May 1997; as of December 1998, construction of the desalination plant had not yet begun, according to Aaron Wolf, Oregon State University, Corvallis, OR, e-mail to author, December 1998.

8. Information Division, "Israeli-Palestinian Interim Agreement Annex III—Protocol Concerning Civil Affairs," Israeli Foreign Ministry, Jerusalem, September 1995.

9. Egypt population from PRB, op. cit. note 6; irrigation expansion plans from official in Egypt's Ministry of Public Works and Water Resources, as quoted in James Dorsey, "Egypt Opens New Canal to Carry Nile Water Beneath Suez to Irrigate Sinai Peninsula," *International Environment Reporter*, 29 October 1997; 8 billion cubic meters (bcm) is author's calculation, assuming an average consumptive use of 8,000 cubic meters

per hectare per year, which may be conservative.

10. Figure of 3.7 million from Z. Abate, "The Integrated Development of Nile Basin Waters," in P.P. Howell and J.A. Allan, eds., *The Nile: Sharing a Scarce Resource* (Cambridge: Cambridge University Press, 1994); Bureau of Reclamation plan from Daniel Hillel, *Rivers of Eden* (New York, Oxford University Press, 1994).

11. Figure of 4 bcm from Hillel, op. cit. note 10; 8 bcm figure from J. A. Allan, "Developing Policies for Harmonised Nile Waters Development and Management," in Howell and Allan, op. cit. note 10.

12. Abate, op. cit. note 10; Amy Dockser Marcus, "Egypt Faces Problem It has Long Dreaded: Less Control of the Nile," *Wall Street Journal*, 22 August 1997.

13. For a similar analysis of this situation, see John Waterbury and Dale Whittington, "Playing Chicken on the Nile? The Implications of Microdam Development in the Ethiopian Highlands and Egypt's New Valley Project," *Natural Resources Forum*, vol. 22, no. 3 (1998).

14. Marcus, op. cit. note 12; Mark Huband, "Egypt a Step Nearer to Taming the Nile," *Financial Times*, 20 February 1998; cost estimate from Gamal Essam El-Din, "Channelling Dissent," *Al Ahram*, 21–28 January 1999, accessed at <imra@netvision. net.il>.

15. Huband, op. cit. note 14.

16. This historical discussion is drawn from Yahia Abdel Mageed, "The Nile Basin: Lessons from the Past," in Asit K. Biswas, ed., *International Waters of the Middle East* (Oxford: Oxford University Press, 1994).

17. Sandra Postel, *Dividing the Waters: Food Security, Ecosystem Health, and the New Politics of Scarcity*, Worldwatch Paper 132 (Washington, DC: Worldwatch Institute, September 1996); Marcus, op. cit. note 12; Mark Huband, "Nile States Look to New Division of Waters," *Financial Times*, 27 February 1998.

18. "Friction Flows over Nile Waters," *World Water and Environmental Engineering*, May 1998.

19. Irrigation expansion plans from Mageed, op. cit. note 16; water requirement range is rough estimation by the author.

20. Sandra Postel, *Last Oasis*, rev. ed. (New York: W.W. Norton & Company, 1997); Hillel, op. cit. note 10; glossy ibis population from Dale Whittington and Elizabeth McClelland, "Opportunities for Regional and International Cooperation in the Nile Basin," unpublished paper, University of North Carolina at Chapel Hill, June 1991.

21. E.A.A. Zaki, "Water Resources Management: Sudan," prepared for World Bank International Workshop on Comprehensive Water Resources Management Policies, Washington DC, 24–28 June 1991;

M.A. Abu-Zeid and M.A. Rady, "Egypt's Water Resources Management and Policies," prepared for ibid.

22. Allan, op. cit. note 11.

23. Biswas et al., op. cit. note 5; Egypt-Sudan incident from McCaffrey, op. cit. note 1.

24. Aaron T. Wolf, "Conflict and Cooperation along International Waterways," *Water Policy,* vol. 1, pp. 251–65 (1998); for a historical perspective, see James L. Wescoat Jr., "Main Currents in Early Multilateral Water Treaties: A Historical-Geographic Perspective, 1648–1948," *Colorado Journal of International Environmental Law and Policy*, vol. 7, no. 1 (1995).

25. United Nations General Assembly, "General Assembly Adopts Convention on Law of Non-Navigational Uses of International Watercourses" (New York: 21 May 1997).

26. United Nations General Assembly, *Convention on the Law of the Non-Navigational Uses of International Watercourses,* New York, 11 April 1997; John R. Crook and Stephen C. McCaffrey, "The United Nations Starts Work on a Watercourses Convention," *American Journal of International Law,* April 1997.

27. Voting pattern from United Nations General Assembly, "Adoption of the Convention on Non-Navigational Uses of International Watercourses," press release (New York: 21 May 1997).

28. Stephen C. McCaffrey, "The Harmon Doctrine One Hundred Years Later: Buried, Not Praised," *Natural Resources Journal,* summer 1996; Demirel quote from John Murray Brown, "Turkey, Syria Set Talks on Euphrates," *Washington Post,* 22 January 1993; quote of Turkey's representative from United Nations General Assembly, op. cit. note 25.

29. Runoff estimates from Luna B. Leopold, *A View of the River* (Cambridge, MA: Harvard University Press, 1994), and from Peter H. Gleick, ed., *Water in Crisis* (New York: Oxford University Press, 1993).

30. Regional Development Administration, *Southeastern Anatolia Project* (Ankara: Republic of Turkey, 1997); $32-billion figure from Regional Development Administration, *Recent Situation in the Southeastern Anatolia Project as of October 1997* (Ankara: Republic of Turkey, 1997); John Kolars, "The Future of the Euphrates River," prepared for World Bank International Workshop on Comprehensive Water Resources Management Policies, Washington, DC, 24–28 June 1991.

31. Water-sharing between Syria and Iraq from John Kolars, "River Advocacy and Return Flow Management on the Euphrates/Firat River: An Important Element in Core-Periphery Relations," prepared for the conference Water: A Trigger for Conflict/A Reason for Cooperation, Indiana

Center on Global Change and World Peace, Bloomington, IN, 7–10 March 1996; Manuel Schiffler, report on the "Interdisciplinary Academic Conference on Water in the Middle East," German Development Institute, Berlin, 17–18 June 1995; John Barham, "Euphrates Power Plant Generates New Tension," *Financial Times*, 15 February 1996; "Syria: Officials Comment on Water Issue with Turkey," Foreign Broadcast Information Service, 19 April 1996, as reported in Arabic by Amir Subur, *Al-Thawrah* (Damascus, Syria), 6 March 1996.

32. Barham, op. cit. note 31; "Turkey, Syria Inch Nearer War," from wire reports, RTD, 5 October 1998.
33. Stephen Kinzer, "Where Kurds Seek a Land, Turks Want the Water," *New York Times*, 28 February 1999; area irrigated and project expenditures from John Kolars, Near East water specialist, Santa Fe, NM, letter to author, January 1999; Regional Development Administration, *Recent Situation*, op. cit. note 30; Stephen Kinzer, "Restoring the Fertile Crescent to its Former Glory," *New York Times*, 29 May 1997.
34. Hillel, op. cit. note 10; Gail Bingham, Aaron Wolf, and Tim Wohlgenant, *Resolving Water Disputes: Conflict and Cooperation in the United States, the Near East, and Asia* (Washington, DC: U.S. Agency for International Development, 1994).
35. Hillel, op. cit. note 10; the Johnston formula allowed Israel to use surplus seasonal flows after the other parties had received their shares, an ambiguous provision that the various parties have interpreted differently.
36. U.S. assistance from Hillel, op. cit. note 10; Jagat S. Mehta, "The Indus Water Treaty: A Case Study in the Resolution of an International River Basin Conflict," *Natural Resources Forum*, vol. 12, no. 1 (1988); Sandra L. Postel, "Resolving Conflict in the Management of Water Resources: Overview of the Problem," in R.G. Sanchez, ed., *Proceedings of the First Biennial Rosenberg International Forum on Water Policy* (Davis, CA: Centers for Water and Wildland Resources, University of California, 1998).
37. Sheila Jones, "When the Ganges Runs Dry," *Financial Times*, 9 May 1994; Zia quoted in Gordon Platt, "India's Control of Ganges River Flow a 'Life and Death' Issue for Bangladesh," *Journal of Commerce*, 26 October 1995.
38. "Treaty Between the Government of the Republic of India and the Government of the People's Republic of Bangladesh on Sharing of the Ganga/Ganges Waters at Farakka," 12 December 1996.
39. Kenneth J. Cooper, "India, Bangladesh Solve Water Dispute," *Washington Post*, 13 December 1996; "Sweeter Waters," *The Economist*, 16 November 1996.

40. "Optimum Use of Ganges Water Aimed at: Hasina," (Bangladesh) *Observer*, March 1998 (day unknown).

41. Biswas et al., op. cit. note 5; evaporation at Aswan from J.A. Allan, "The Nile Basin: Water Management Strategies," in Howell and Allan, op. cit. note 10; possible savings from Hillel, op. cit. note 10; last example is author's calculation and assumes water consumption of 1,250 cubic meters per ton of grain and net grain imports averaged over 1994–96, from U.S. Department of Agriculture, *Production, Supply, and Distribution,* electronic database, Washington, DC, February 1997.

42. Hillel, op. cit. note 10.

43. Aaron T. Wolf, "Criteria for Equitable Allocations: The Heart of International Water Conflict," *Natural Resources Forum*, forthcoming.

44. Short-term incentive of political leaders from Biswas et al., op. cit. note 5.

45. Sander Thoenes, "Central Asians Reach Common Ground over Water," *Financial Times*, 9 April 1996.

46. Gretta Goldenman, "International River Agreements in the Context of Climatic Changes," Pacific Institute for Studies in Development, Environment, and Security, Berkeley, CA, 1989.

47. Sandra L. Postel, Jason I. Morrison, and Peter H. Gleick, "Allocating Fresh Water to Aquatic Ecosystems: The Case of the Colorado River Delta," *Water International*, September 1998; Jason I. Morrison, Sandra L. Postel, and Peter H. Gleick, *The Sustainable Use of Water in the Lower Colorado River Basin* (Oakland, CA: The Pacific Institute, 1996).

48. Warren Christopher, "American Diplomacy and the Global Environmental Challenges of the Twenty-first Century," address at Stanford University, Stanford, CA, 9 April 1996.

49. Amy Dockser Marcus, "Greenpolitik: Threats to Environment Provoke A New Security Agenda," *Wall Street Journal*, 20 November 1997; Strobe Talbott, "The Global Environment and the National Interest," address to the Foreign Service Institute, Washington, DC, 10 September 1996.

Chapter 8. The Productivity Frontier

1. Grainland productivity from U.S. Department of Agriculture (USDA), *Production, Supply, and Distribution*, electronic database, Washington, DC, updated February 1999; rise in irrigation water use from I. A. Shiklomanov, "Assessment of Water Resources and Water Availability in the World" (St. Petersburg, Russia: State Hydrological Institute, February 1996).

2. David Seckler, *The New Era of Water Resources Management: From "Dry"*

to "Wet" Water Savings (Washington, DC: Consultative Group on International Agricultural Research (CGIAR), 1996).

3. David Molden, "Accounting for Water Use and Productivity," System-Wide Initiative for Water Management, Paper No. 1 (Colombo, Sri Lanka: International Water Management Institute (IWMI), 1997).

4. Developing-country projection from Mark W. Rosegrant and Claudia Ringler, "Impact on Food Security and Rural Development of Reallocating Water from Agriculture for Other Uses," paper prepared for Harare Expert Group Meeting on Strategic Approaches to Freshwater Management, Harare, Zimbabwe, 28–31 January 1998.

5. For a good description of drip irrigation's development, see pages 221–25 of Daniel Hillel, *Rivers of Eden* (New York: Oxford University Press, 1994).

6. Sandra Postel, *Conserving Water: The Untapped Alternative*, Worldwatch Paper 67 (Washington, DC: Worldwatch Institute, September 1985); Sandra Postel, *Last Oasis*, rev. ed. (New York: W.W. Norton & Company, 1997); Dale A. Bucks, "Historical Developments in Microirrigation," in Freddie R. Lamm, ed., *Microirrigation for a Changing World*, Proceedings of the Fifth International Microirrigation Congress (St. Joseph, MI: American Society of Agricultural Engineers, 1995); Dale A. Bucks, USDA, Agricultural Research Service, Beltsville, MD, discussion with author, 1 July 1998.

7. S.K. Suryawanshi, "Success of Drip in India: An Example to the Third World," in Lamm, op. cit. note 6; World Bank, *Gains That Might be Made from Water Conservation in the Middle East* (Washington, DC: 1993); average capital costs from Amy Vickers, *Handbook of Water Use and Conservation* (Boca Raton, FL: Lewis Publishers, in press).

8. Indian National Committee on Irrigation and Drainage, *Drip Irrigation in India* (New Delhi: 1994); R.K. Sivanappan, chair of the Centre of Agricultural Rural Development and Environmental Studies in Coimbatore, discussions with author during field visits in Madhya Pradesh, India, January 1998.

9. Indian National Committee, op. cit. note 8; R.K. Sivanappan, "Prospects of Micro-Irrigation in India," *Irrigation and Drainage Systems*, vol. 8, pp. 49–58 (1994).

10. Joseph C. Henggeler, "A History of Drip-Irrigated Cotton in Texas," in Lamm, op. cit. note 6.

11. Ibid.; description of Sundance Farm operation from Tom Murphy, "Low-Impact Agricultural Irrigation," *Irrigation Journal*, March 1995.

12. Henggeler, op. cit. note 10.

13. Lynn Moseley, "Deaf Smith County Producer Pleased with Drip Irriga-

tion Results," *The Cross Section* (High Plains Underground Water Conservation District), March 1998.

14. Sugarcane's water share in Maharashtra from World Bank, *India: Irrigation Sector Review*, vol. 1 (Washington, DC: 1991); nationwide sugarcane area from H.S. Chauhan, "Issues of Standardisation and Scope of Drip Irrigation in India," in Lamm, op. cit. note 6.

15. S.S. Magar, "Adoption of Microirrigation Technology in Sugarcane (Saccharum Officinarum L.) on Vertisols in Semi Arid Climate," in Lamm, op. cit. note 6; Indian National Committee, op. cit. note 8; Hawaii from World Bank, op. cit. note 7.

16. Vickers, op. cit. note 7.

17. Paul Polak, Bob Nanes, and Deepak Adhikari, "A Low Cost Drip Irrigation System for Small Farmers in Developing Countries," *Journal of the American Water Resources Association*, February 1997.

18. Author's travels and field interviews in the lower Himalayas of Himachel Pradesh, India, January 1998; installations in India and Nepal from Jeff Saussier, International Development Enterprises (IDE), Denver, CO, letter to author, December 1998.

19. Drip initiative and India demonstration project from Paul Polak, President, IDE, Denver, CO, discussions with author, various dates in 1998.

20. California Department of Water Resources, "CIMIS: Fifteen Years of Growth and a Promising Future" (Sacramento, CA: December 1997).

21. Ibid.; the source does not distinguish the portion of water savings from reduced evapotranspiration.

22. Vickers, op. cit. note 7.

23. Mary DeSena, "Satellites Take the Guess Work Out of Irrigation," *U.S. Water News*, July 1998; Fred Ziari, president, IRZ Consulting, Hermiston, OR, discussion with author, 6 July 1998.

24. Vickers, op. cit. note 7.

25. CGIAR, "The World Water and Climate Atlas for Agriculture: A New Technology," press release (Washington, DC: 16 March 1997); the atlas is available at <http://www.cgiar.org/iwmi>.

26. "Developing a Digital Atlas of the World's Water Balance," *New Waves* (Texas Water Resources Institute), June 1998; this atlas is available at <http://www.ce.utexas.edu/prof/maidment/atlas/ atlas.htm>.

27. San Joaquin Valley Drainage Implementation Program, *Drainage Management in the San Joaquin Valley: A Status Report* (Sacramento, CA: California Department of Water Resources, 1998).

28. Ibid.

29. Ramesh Bhatia, Rita Cestti, and James Winpenny, *Water Conservation and Reallocation: Best Practice Cases in Improving Economic Efficiency*

and Environmental Quality (Washington, DC: World Bank, 1995).

30. Ibid.

31. Ibid.

32. Lynn Moseley, "Urban, Agricultural Water Conservation Tips Offered by District," *The Cross Section* (High Plains Underground Water Conservation District), June 1996.

33. Ibid.; costs and payback from Vickers, op. cit. note 7.

34. Tamara Daniel, "TAES Research Focuses on Center Pivot Irrigation Management Decisions," *The Cross Section* (High Plains Underground Water Conservation District), June 1995; payback period from Postel, *Last Oasis*, op. cit. note 6; corn and cotton yield increases and sprinkler coverage from Ken Carver, Assistant Manager, High Plains Underground Water Conservation District No. 1, Lubbock, TX, discussion with author, 21 July 1998.

35. "Economic Benefits of Water Conservation Techniques Discussed During Quarterly Ogallala Regional Water Management Plan Meeting," *The Cross Section* (High Plains Underground Water Conservation District), July 1996.

36. Carver, op. cit. note 34; water table declines from "Ground Water Levels Decline 0.34 of a Foot within District in 1997," *The Cross Section* (High Plains Underground Water Conservation District), April 1998.

37. Lynn Moseley, "District Tests Use of PET Data to Improve Irrigation Scheduling," *The Cross Section* (High Plains Underground Water Conservation District), January 1997; Carver, op. cit. note 34.

38. Figure of 100 million from G.A. Cornish, *Modern Irrigation Technologies for Smallholders in Developing Countries* (Wallingford, U.K.: HR Wallingford, Ltd., 1997); 90-percent figure from L.C. Guerra et al., *Producing More Rice with Less Water from Irrigated Systems* (Colombo, Sri Lanka: IWMI, 1998).

39. Sri Lanka example from J.S. Wallace and C.H. Batchelor, "Managing Water Resources for Crop Production," *Philosophical Transactions of the Royal Society of London: Biological Sciences,* vol. 352, pp. 937–47 (1997).

40. Guerra et al., op. cit. note 38.

41. Ibid.

42. Ramesh Bhatia, Upali Amerasinghe, and K.A.U.S. Imbulana, "Productivity and Profitability of Paddy Production in the Muda Scheme, Malaysia," *Water Resources Development,* vol. 11, no. 1 (1995).

43. Ibid.

44. Gary N. McCauley, Lloyd R. Hossner, and Doublas M. Nesmith, "Sprinkler Irrigation as an Energy and Water Saving Approach to Rice Production and Management of Riceland Pests," Texas Water Resources

Institute, Texas A&M University System, College Station, TX, March 1985.

45. Current harvest index and potential for increases from Charles C. Mann, "Crop Scientists Seek a New Revolution," *Science*, 15 January 1999.

46. For an excellent review, see T.R. Sinclair, C.B. Tanner, and J.M. Bennett, "Water-Use Efficiency in Crop Production," *BioScience*, January 1984; Mann, op. cit. note 45; F.W.T. Penning deVries, H. Van Keulen, and R. Rabbinge, "Natural Resources and Limits of Food Production in 2040," in J. Bouma et al., eds., *Eco-Regional Approaches for Sustainable Land Use and Food Production* (Dordrecht, Netherlands: Kluwer Academic Publishers, 1995).

47. Pamela C. Ronald, "Making Rice Disease-Resistant," *Scientific American*, November 1997.

48. Size of wheat genome and mapping technique from ibid.

49. National Research Council, *Lost Crops of Africa, Vol. I: Grains* (Washington, DC: National Academy Press, 1996).

50. Israel's reuse from Hillel Shuval, presentation at the Stockholm Environment Institute/United Nations Workshop on Freshwater Resources, New York, 18–19 May 1996; Tunisia example from World Bank, *From Scarcity to Security: Averting a Water Crisis in the Middle East and North Africa* (Washington, DC: 1995); Tenerife example from Ionics, Inc., *Ionics 1997 Annual Report* (Watertown, MA: 1998).

51. Guerra et al., op. cit. note 38.

52. Aldo Leopold, *A Sand County Almanac* (New York: Oxford University Press, 1949).

53. Species from E.P. Glenn et al., "Cienega de Santa Clara: Endangered Wetland in the Colorado River Delta, Sonora, Mexico," *Natural Resources Journal*, Fall 1992; endangered species from E.P. Glenn et al., "Effects of Water Management on the Wetlands of the Colorado River Delta, Mexico," *Conservation Biology*, August 1996.

54. Sandra L. Postel, Jason I. Morrison, and Peter H. Gleick, "Allocating Fresh Water to Aquatic Ecosystems: The Case of the Colorado River Delta," *Water International*, September 1998.

55. Jerry Emory, "Just Add Water," *Nature Conservancy*, November-December 1994.

56. For a short history of the partnership, see Marc Reisner, "Deconstructing the Age of Dams: California Rice Farmers, of All People, Lead the Way," *High Country News*, 27 October 1997.

57. Emory, op. cit. note 55.

Chapter 9. Thinking Big About Small-Scale Irrigation

1. Paul Polak, "Dismantling the Barriers to the Green Revolution for Small Farmers" (Denver, CO: International Development Enterprises (IDE), March 1996); World Bank, *Food Security for the World,* Statement prepared for the World Food Summit, 12 November 1996.
2. Some 840 million people remain malnourished, according to U.N. Food and Agriculture Organization (FAO), *The Sixth World Food Survey* (Rome: 1996).
3. Data for Figure 9–1 from FAO, *Water Development for Food Security* (Rome: March 1995).
4. Alastair Orr, A.S.M. Nazrul Islam, and Gunnar Barnes, *The Treadle Pump: Manual Irrigation for Small Farmers in Bangladesh* (Dhaka, Bangladesh: Rangpur Dinajpur Rural Service, 1991).
5. Author's field visits and discussions with farmers in Brahmanbaria, Bangladesh, January 1998.
6. Paul Polak, "Tripling the Small Farmer Harvest: A Solution for Rural Hunger" (Denver, CO: IDE, January 1998); small farmers will typically be attracted to an innovation that increases returns two- to threefold, according to G.A. Cornish, *Modern Irrigation Technologies for Smallholders in Developing Countries* (Wallingford, U.K.: H.R. Wallingford, 1997).
7. Author's field visits in Bangladesh, January 1998; 1.2 million figure from Len Jornlin, IDE Country Director, Dhaka, Bangladesh, e-mail to author, 23 August 1998; economic impact based on estimated net returns of $100 per pump and a per-pump multiplier of 2.5; growth in irrigated area from FAO, *1996 Production Yearbook* (Rome: 1997).
8. Background on IDE from promotional material, IDE, Denver, CO; role of IDE from Orr, Nazrul Islam, and Barnes, op. cit. note 4; film from Polak, op. cit. note 6.
9. Number of manufacturers, dealers, and installers as of October 1998 from Jeff Saussier, IDE, Denver, CO, letter to author, December 1998; potential market from Paul Polak, President, IDE, discussion with author during trip to Bangladesh, January 1998.
10. Polak, op. cit. note 6; problem posed by deep tubewells from Ruth Meinzen-Dick, Research Fellow, International Food Policy Research Institute (IFPRI), Washington, DC, e-mail to author, December 1998.
11. Narendra P. Sharma et al., *African Water Resources* (Washington, DC: World Bank, 1996); one third lacking food from U.N. Development Programme, *Human Development Report 1996* (New York: 1996); import needs from Mark W. Rosegrant and Nicostrato D. Perez, "Water

Resources Development in Africa: A Review and Synthesis of Issues, Potentials, and Strategies for the Future," Discussion Paper No. 28 (Washington, DC: Environment and Production Technology Division, IFPRI, September 1997); Per Pinstrup-Anderson, Rajul Pandya-Lorch, and Mark W. Rosegrant, *The World Food Situation: Recent Developments, Emerging Issues, and Long-Term Prospects* (Washington, DC: IFPRI, 1997).

12. African ratios of runoff-to-precipitation from FAO, *Irrigation in Africa in Figures,* Water Reports No. 7 (Rome: 1995); global figure from Sandra L. Postel, Gretchen C. Daily, and Paul R. Ehrlich, "Human Appropriation of Renewable Fresh Water," *Science*, 9 February 1996; Sharma et al., op. cit. note 11.

13. Rosegrant and Perez, op. cit. note 11; Sharma et al., op. cit. note 11.

14. Cropland that is irrigated from FAO, op. cit. note 7; FAO, op. cit. note 12.

15. A. Kandiah, "Summary of Findings of Missions in Selected Countries in East and Southern Africa," in FAO, *Irrigation Technology Transfer in Support of Food Security*, Proceedings of a Subregional Workshop, Harare, Zimbabwe, 14–17 April 1997, Water Reports No. 14 (Rome: 1997).

16. For an excellent treatment of indigenous water management in Africa, see W.M. Adams, *Wasting the Rain: Rivers, People, and Planning in Africa* (London: Earthscan, 1992); for an instructive case study, see W.M. Adams, T. Potkanski, and J.E.G. Sutton, "Indigenous Farmer-Managed Irrigation in Sonjo, Tanzania," *The Geographical Journal*, March 1994.

17. Definition quoted in Adams, op. cit. note 16; FAO, op. cit. note 12.

18. Adams, op. cit. note 16.

19. Marc Andreini, "*Bani* Irrigation: An Alternative Water Use," in M. Rukuni et al., eds., *Irrigation Performance in Zimbabwe*, Proceedings of Workshops in Harare and Juliasdale, Zimbabwe (University of Zimbabwe: 1994); Meinzen-Dick, op. cit. note 10.

20. Zambia figure from Keith Frausto, "The Introduction of the Treadle Pump in Zambia," *IDEAS* (newsletter of IDE, Denver, CO), summer 1997; Zimbabwe *dambo* figures from K. Palanisami, "Economics of Irrigation Technology Transfer and Adoption," in FAO, op. cit. note 15; Zimbabwe total irrigated area from FAO, op. cit. note 7; Rukuni et al., op. cit. note 19.

21. Randall Purcell, "Potential for Small-Scale Irrigation in Sub-Saharan Africa: The Kenyan Example," in FAO, op. cit. note 15.

22. Sharma et al., op. cit. note 11.

23. Ibid.; falling water levels from Adams, op. cit. note 16.

24. Ellen P. Brown and Robert Nooter, "Successful Small-Scale Irrigation in the Sahel" (draft) (Washington, DC: World Bank, September 1991).

25. Ibid.
26. Presence of cultivars from Thayer Scudder, "The Need and Justification for Maintaining Transboundary Flood Regimes: The Africa Case," *Natural Resources Journal*, winter 1991; quote appears in Michael M. Horowitz and Muneera Salem-Murdock, "River Basin Development Policy, Women, and Children: A Case Study from the Senegal River Valley," in Filomina Chioma Steady, ed., *Women and Children First: Environment, Poverty and Sustainable Development* (Rochester, VT: Schenkman Books, Inc.: 1993).
27. Area cultivated from Scudder, op. cit. note 26.
28. Ibid.; Horowitz and Salem-Murdock, op. cit. note 26.
29. Michael M. Horowitz, "The Management of an African River Basin: Alternative Scenarios for Environmentally Sustainable Economic Development and Poverty Alleviation," in Proceedings of the International UNESCO Symposium, *Water Resources Planning in a Changing World* (Karlsruhe, Germany: Bundesanstalt für Gewässerkunde, 1994).
30. Michael M. Horowitz and Muneera Salem-Murdock, "Development-Induced Food Insecurity in the Middle Senegal Valley," *GeoJournal*, vol. 30, no. 2 (1993); quote from Horowitz, op. cit. note 29.
31. Horowitz, op. cit. note 29; farmers' union demands from Lori Pottinger, "Dissent Grows over Senegal River Valley Dams," *World Rivers Review*, February 1998.
32. IDE, "Low Cost Drip Irrigation Fact Sheet," undated, <http://www.ideorg.org>.
33. Ibid.; author's visit to drip irrigation sites in Himachel Pradesh, India, January 1998.
34. Figure of 10 percent based on I.A. Shiklomanov, "Assessment of Water Resources and Water Availability in the World" (St. Petersburg, Russia: State Hydrological Institute, February 1996), who estimates that sprinklers irrigate about 20 million hectares—8 percent of 1995 world irrigated area.
35. Author's field visits to IDE demonstration plots in Madhya Pradesh, India, January 1998; Saussier, op. cit. note 9.
36. Partitioning of rainfall from J.S. Wallace, Institute of Hydrology, Wallingford, U.K., in presentation at the Ninth World Water Congress, International Water Resources Association, Montreal, Canada, 1–6 September 1997.
37. Anil Agarwal and Sunita Narain, eds., *Dying Wisdom* (New Delhi: Centre for Science and Environment, 1997).
38. Ibid.; Ganesh Pangare, "Traditional Water Harvesting on Way Out," *Economic and Political Weekly*, 7–14 March 1992.

39. Agarwal and Narain, op. cit. note 37.

40. Ibid.

41. Number of tanks and related irrigated area from R.K. Sivanappan, "Technologies for Water Harvesting and Soil Moisture Conservation in Small Watersheds for Small-Scale Irrigation," in FAO, op. cit. note 15; Agarwal and Narain, op. cit. note 37.

42. T.M Mukundan, "Tamil Nadu: Tank Traditions," in Agarwal and Narain, op. cit. note 37; occasional assistance from kings or warlords from Meinzen-Dick, op. cit. note 10.

43. Mukundan, op. cit. note 42.

44. Ibid.

45. Agarwal and Narain, op. cit. note 37; Uma Shankari, "Tanks: Major Problems in Minor Irrigation," *Economic and Political Weekly*, 28 September 1991.

46. Yield increase from A.M. Michael, "Raising Yield in Rainfed Lands: Stress on Water Management," in The Hindu, *Survey of Indian Agriculture 1990* (Madras, India: M/s. Kasturi & Sons Ltd., undated).

47. Agarwal and Narain, op. cit. note 37; Anil Agarwal and Sunita Narain, "Dying Wisdom: The Decline and Revival of Traditional Water Harvesting Systems in India," *The Ecologist*, May-June 1997.

48. Sandra Postel, *Last Oasis*, rev. ed. (New York: W.W. Norton & Company, 1997); author's visit to Negev Desert and Jacob Blaustein Institute for Desert Research, Ben Gurion University of the Negev, Sde Boqer Campus, Israel, March 1992.

49. Sivanappan, op. cit. note 41.

50. Author's field visit, Madhya Pradesh, India, January 1998.

51. Werner Hunziker, Head, Natural Resource Management, Swiss Agency for Development and Cooperation, Embassy of Switzerland, New Delhi, discussion with author, 22 January 1998.

Chapter 10. The Players and the Rules

1. India figures from A. Vaidyanathan, "Second India Series Revisited: Food and Agriculture," prepared for the World Resources Institute (WRI), Washington, DC, n.d.; Tunisia and Jordan examples from World Bank, *From Scarcity to Security: Averting a Water Crisis in the Middle East and North Africa* (Washington, DC: 1995); Norman Myers, "Perverse Subsidies: Their Nature, Scale and Impacts," a report to the MacArthur Foundation, Chicago, IL, October 1997.

2. For a good discussion of rent-seeking in irrigation, see Robert Repetto, *Skimming the Water* (Washington, DC: WRI, 1986); Richard W. Wahl,

Markets for Federal Water (Washington, DC: Resources for the Future, 1989).

3. Repetto, op. cit. note 2; Congressional Budget Office cited in Subcommittee on Oversight and Investigations, Committee on Natural Resources, U.S. House of Representatives, *Taking from the Taxpayer: Public Subsidies for Natural Resource Development* (Washington, DC: 1994).

4. Subcommittee on Oversight and Investigations, op. cit. note 3.

5. Terry L. Anderson and Pamela S. Snyder, *Priming the Invisible Pump* (Bozeman, MT: PERC Policy Center, February 1997).

6. Robert Chambers, *Managing Canal Irrigation: Practical Analysis from South Asia* (Cambridge, U.K.: Cambridge University Press, 1988).

7. Ibid.

8. Figure of $31 billion from William I. Jones, *The World Bank and Irrigation* (Washington, DC: World Bank, 1995).

9. Chambers, op. cit. note 6; 44-percent figure from Patrick McCully, *Silenced Rivers: The Ecology and Politics of Large Dams* (London: Zed Books, 1996).

10. Canal example described in Elinor Ostrom, *Crafting Institutions for Self-Governing Irrigation Systems* (San Francisco: ICS Press, 1992); quote from Chambers, op. cit. note 6.

11. Margreet Zwarteveen, *Linking Women to the Main Canal* (London: International Institute for Environment and Development, 1995).

12. World Bank assessment from Jones, op. cit. note 8.

13. Ostrom, op. cit. note 10.

14. C.J. Perry, Michael Rock, and D. Seckler, "Water as an Economic Good: A Solution, or a Problem?" Research Report 14 (Colombo, Sri Lanka: International Water Management Institute (IWMI), 1997); Jones, op. cit. note 8.

15. Irrigated area in Punjab from Ruth Meinzen-Dick, *Groundwater Markets in Pakistan* (Washington, DC: International Food Policy Research Institute, 1996).

16. C.J. Perry and S.G. Narayanamurthy, "Farmer Response to Rationed and Uncertain Irrigation Supplies," Research Report 24 (Colombo, Sri Lanka: IWMI, 1998).

17. For a discussion of demand-driven versus supply-driven systems, see Jones, op. cit. note 8.

18. Conditions needed for warabandi from Chambers, op. cit. note 6.

19. Pricing structure from Michael R. Moore and Ariel Dinar, "Water and Land as Quantity-Rationed Inputs in California Agriculture: Empirical Tests and Water Policy Implications," *Land Economics*, November 1995.

---

20. Anderson and Snyder, op. cit. note 5.
21. Ibid.
22. "Buying a Gulp of the Colorado," *The Economist*, 24 January 1998; Christopher Parkes, "Water Wars Ebb Away in the West," *Financial Times*, 8 January 1998; Todd S. Purdum, "U.S. Acts to Meet Water Needs of Growing States in the West," *New York Times*, 19 December 1997.
23. Lawrence J. MacDonnell et al., "Using Water Banks to Promote More Flexible Water Use," Report to the U.S. Geological Survey (Boulder, CO: Natural Resources Law Center, August 1994); Anderson and Snyder, op. cit. note 5; California Department of Water Resources, *The California Water Plan Update,* Public Review Draft, Bulletin 160–98, vol. 1 (Sacramento, CA: January 1998).
24. Mark W. Rosegrant and Hans P. Binswanger, "Markets in Tradable Water Rights: Potential for Efficiency Gains in Developing Country Water Resource Allocation," *World Development,* vol. 22, no. 11 (1994); Pakistan figure from Mateen Thobani, "Tradable Property Rights to Water," FPD Note No. 34 (Washington, DC: World Bank, February 1995).
25. Meinzen-Dick, op. cit. note 15.
26. Tushaar Shah, *Groundwater Markets and Irrigation Development* (Bombay: Oxford University Press, 1993); Tushaar Shah and Vishwa Ballabh, "Water Markets in North Bihar," *Economic and Political Weekly*, 27 December 1997.
27. Shah and Ballabh, op. cit. note 26.
28. Douglas L. Vermillion, *Impacts of Irrigation Management Transfer* (Colombo, Sri Lanka: IWMI, 1997).
29. Jones, op. cit. note 8; Ostrom, op. cit. note 10.
30. Ostrom, op. cit. note 10.
31. Ibid.; Harald D. Frederiksen, Jeremy Berkoff, and William Barber, *Water Resources Management in Asia* (Washington, DC: World Bank, 1993).
32. The study is that of Joost Oorthuizen and Wim H. Kloezen, "The Other Side of the Coin: A Case Study on the Impact of Financial Autonomy on Irrigation Management Performance in the Philippines," *Irrigation and Drainage Systems*, vol. 9, pp. 15–37 (1995), as described in Vermillion, op cit. note 28.
33. Study reported in Jones, op. cit. note 8.
34. Sam H. Johnson III, "Irrigation Management Transfer: Decentralizing Public Irrigation in Mexico," *Water International*, vol. 22, pp. 159–67 (1997).
35. Ibid.
36. Ibid.; for Mexican water law and nature of irrigation concessions, see

Cecilia M. Gorriz, Ashok Subramanian, and José Simas, *Irrigation Management Transfer in Mexico* (Washington, DC: World Bank, 1995).

37. Johnson, op. cit. note 34.

38. Wim H. Kloezen, Carlos Garces-Restrepo, and Sam H. Johnson III, *Impact Assessment of Irrigation Management Transfer in the Alto Rio Lerma Irrigation District, Mexico* (Colombo, Sri Lanka: IWMI, 1997).

39. Johnson, op. cit. note 34; for more on Turkey's transfer program, see Mark Svendsen and Gladys Nott, "Irrigation Management Transfer in Turkey: Early Experience with a National Program under Rapid Implementation" (Colombo, Sri Lanka: IWMI, 1997).

40. Garrett Hardin, "The Tragedy of the Commons," *Science*, 13 December 1968.

41. Ostrom, op. cit. note 10.

42. Ariel Dinar, Mark W. Rosegrant, and Ruth Meinzen-Dick, *Water Allocation Mechanisms: Principles and Examples*, Policy Research Working Paper 1779 (Washington, DC: World Bank, 1997).

43. A strong case for negotiated allocation processes that recognize customary water rights is made by Bryan Bruns and Ruth S. Meinzen-Dick, eds., *Negotiating Water Rights* (New Delhi: Sage Publishers, forthcoming).

44. V. Narain, "Towards a New Groundwater Institution for India," *Water Policy*, vol.1, pp. 357–65 (1998).

45. Ibid.; Marcus Moench, "Allocating the Common Heritage: Debates over Water Rights and Governance Structures in India," *Economic and Political Weekly*, 27 June 1998.

46. Sandra Postel, *Last Oasis*, rev. ed. (New York: W.W. Norton & Company, 1997); suggestions for India from Narain, op. cit. note 45, and from Moench, op. cit. note 46.

47. J. Eduardo Mestre R., "Integrated Approach to River Basin Management: Lerma-Chapala Case Study—Attributions and Experiences in Water Management in Mexico," *Water International*, September 1997.

Chapter 11. Listening to Ozymandias

1. Peter Christensen, *The Decline of Iranshahr* (Copenhagen, Denmark: Museum Tusculanum Press, University of Copenhagen, 1993).

2. Jared Diamond, *Guns, Germs, and Steel: The Fates of Societies* (New York: W.W. Norton & Company, 1997).

3. United Nations, *World Population Prospects 1950–2050—The 1998 Revision* (New York: December 1998).

4. Figure of $4.9 trillion from Robert Costanza et al., "The Value of the

World's Ecosystem Services and Natural Capital," *Nature*, 15 May 1997; Sandra Postel and Stephen Carpenter, "Freshwater Ecosystem Services," in Gretchen C. Daily, ed., *Nature's Services* (Washington, DC: Island Press, 1997); gross world product of $39 trillion from Worldwatch update of Angus Maddison, *Monitoring the World Economy 1820–1992* (Paris: Organisation for Economic Co-operation and Development, 1995).

5. Evan Eisenberg, *The Ecology of Eden* (New York: Alfred A. Knopf, 1998).

6. Water requirements of U.S. diet from U.N. Food and Agriculture Organization, *World Food Summit—Technical Background Documents*, Vol. 2 (Rome: 1996).

7. Pork consumption figures from U.S. Department of Agriculture, Foreign Agricultural Service, "Livestock and Poultry: World Markets and Trade," Washington, DC, October 1998; China's share of world population from Population Reference Bureau, *1998 World Population Data Sheet*, wallchart (Washington, DC: 1998).

8. Aldo Leopold, *A Sand County Almanac* (New York: Oxford University Press, 1949).

9. Appropriation of fresh water from Sandra L. Postel, Gretchen C. Daily, and Paul R. Ehrlich, "Human Appropriation of Renewable Fresh Water," *Science*, 9 February 1996; appropriation of photosynthetic product from Peter M. Vitousek et al., "Human Appropriation of the Products of Photosynthesis," *BioScience*, June 1986.

10. United Nations, op. cit. note 3.

INDEX

ABOUT THE AUTHOR

SANDRA POSTEL is Director of the Global Water Policy Project in Amherst, Massachusetts, where her research focuses on international water issues and strategies. From 1988 until 1994, she served as Vice President for Research at the Worldwatch Institute, a private nonprofit research organization with which she remains affiliated as Senior Fellow. In 1995, Ms. Postel also became a Pew Fellow in Conservation and the Environment.

Ms. Postel has published widely in scholarly and popular publications, and has lectured at many universities, including Stanford, Harvard, Duke, MIT, and Yale. For two years, she taught at Tufts University as Adjunct Professor of International Environmental Policy. She has served on the Board of Directors of the International Water Resources Association and the World Future Society, as an advisor to the Global 2000 program founded by President Jimmy Carter, and as a consultant to the United Nations Development Programme and the World Bank. She currently serves on a number of journal editorial boards and as a senior advisor to the World Commission on Water.

Ms. Postel's earlier book, *Last Oasis*, appears in eight languages and was chosen by *Choice* magazine as one of the outstanding academic books of 1993. *Last Oasis* was also the basis for a PBS documentary that first aired in 1997.